INTERDISCIPLINARITY IN THE SCIENCES

Nonlinear Dynamics, Chaos and Complexity

INTERDISCIPLINARITY IN THE SCIENCES

Nonlinear Dynamics, Chaos and Complexity

Miguel A F Sanjuán

Universidad Rey Juan Carlos, Madrid, Spain

NEW JERSEY · LONDON · SINGAPORE · BEIJING · SHANGHAI · HONG KONG · TAIPEI · CHENNAI · TOKYO

Published by

World Scientific Publishing Co. Pte. Ltd.

5 Toh Tuck Link, Singapore 596224

USA office: 27 Warren Street, Suite 401-402, Hackensack, NJ 07601

UK office: 57 Shelton Street, Covent Garden, London WC2H 9HE

British Library Cataloguing-in-Publication Data
A catalogue record for this book is available from the British Library.

INTERDISCIPLINARITY IN THE SCIENCES
Nonlinear Dynamics, Chaos and Complexity

ISBN 978-981-98-1508-1 (hardcover)
ISBN 978-981-98-1509-8 (ebook for institutions)
ISBN 978-981-98-1510-4 (ebook for individuals)

For any available supplementary material, please visit
https://www.worldscientific.com/worldscibooks/10.1142/14364#t=suppl

Desk Editor: Rhaimie B Wahap

Printed in Singapore

To Professor James A. Yorke, who perhaps unknowingly has been a guiding light and a constant inspiration throughout my journey in Nonlinear Dynamics, Chaos, and Complex Systems.

Endorsements

"The book explores the interdisciplinary nature of modern research through the unifying framework of Nonlinear Dynamics, Chaos, and Complex Systems. It vividly illustrates how these fundamental concepts have fertilized advances across mathematics, physics, chemistry, biology, geology, and beyond."

Guanrong (Ron) Chen

City University of Hong Kong, P R China

"Miguel Sanjuán's fresh and original view on the history guides us in a fun and deep manner to the recent evolution as well as perspectives and future potentials of nonlinear sciences."

Juergen Kurths

Potsdam Institute for Climate Impact Research (PIK), Germany

and

Humboldt University of Berlin, Germany

"Interdisciplinary perspectives on our complex world through the power of nonlinear science, which goes beyond the walls of different disciplines."

Kazuyuki Aihara

International Research Center for Neurointelligence (IRCN), The University of Tokyo, Japan

"In this masterfully crafted work, Miguel A. F. Sanjuán distills the rich history, foundational ideas, and modern applications of nonlinear dynamics, chaos, and complexity into an inspiring narrative that transcends disciplinary boundaries. Bridging physics, life sciences, mathematics, and beyond, this book not only illuminates the conceptual beauty of complex systems but also demonstrates their unifying power in advancing truly interdisciplinary science. A must-read for researchers, educators, and all who seek to understand the nonlinear world we live in."

Dimitri Volchenkov

Texas Tech University, USA

"This most timely book covers credibly, without mathematical equations, the historical developments, foundational contributions by Nobel laureates and diverse applications of Nonlinear Dynamics, Chaos, and Complexity. It foregrounds the demands for teaching complexity in college level, offers the strategic planning of curriculum design and stresses the needs for interdepartmental and

international collaborations which are not touched by other books on complex phenomena."

S. Rajasekar

Bharathidasan University, India

"A fascinating and highly absorbing historical introduction to nonintegrable nonlinear, namely chaotic, dynamical systems. It introduces in a masterful way the notions of emergence, complexity and their diverse applications in physics, chemistry, biology, engineering, social and earth sciences as well as the interdisciplinary nature of the subject. It also provides clear ideas of introducing relevant courses in scientific curriculum."

M. Lakshmanan

Bharathidasan University, India

"This book provides a bird's eye view of the remarkable history and achievements of the sciences described. The presentation is inspiring and captivating. A highly recommendable book to enter in these fascinating subjects, with relevant and useful references to all topics discussed that will allow the reader to find more detailed descriptions of the broad variety of topics covered."

Luis Álvarez-Gaumé

Stony Brook University, USA

"Prof. Sanjuán's acceptance speech at the Royal Academy of Sciences brought a breath of fresh air to the Institution, an ode to chaos and complexity. "It has been the great triumph of the sciences to find consistent means of studying phenomena hidden by both space and time, overcoming the limits of cognition and material culture" writes Prof. David C. Krakauer, current director of the Santa Fe Institute, sanctum of complexity. The unpredictable and the complex are treated with such mastery that the exact, the physical, and the natural appear as a normal mixture that permeates all areas of knowledge.

Pedro R. García Barreno

Royal Spanish Academy & Royal Academy of Sciences of Spain, Spain

"This terrifically readable account of the emergence of complexity science weaves together the central threads of determinism, interdisciplinarity and emergence. It is a personal account of the key figures in the history of nonlinear science and complexity and draws in the author's invaluable and perceptive insight into the evolution of modern science and mathematics. Well written, warm, and wonderfully insightful tour of the development of nonlinear science. One gets a clear sense of the gradual building of both a new kind of science, and of the recognition for the pioneers in that science."

Michael Small

The University of Western Australia, Australia

Prologue

The main objective of this book, as its title suggests, is to emphasize the importance of interdisciplinarity in scientific research through the framework of Nonlinear Dynamics, Chaos, and Complex Systems. It is based on the speech I delivered at my induction ceremony into the Royal Academy of Sciences of Spain on June 27, 2024. The primary aim of that address was to offer an overview of the scientific field to which I have dedicated my career[1], underscoring the inherently interdisciplinary nature, broad scientific scope, and international impact of Nonlinear Dynamics, Chaos, and Complex Systems.

While preparing my inaugural speech, I was honored to receive an invitation from Lakshmi Narayanan of World Scientific Publishing, who encouraged expanding my address into a book written in English. This proposal arose from the significant interest the speech had generated and the perceived value of offering a comprehensive

---

[1] My personal intellectual journey toward nonlinear dynamics is described in detail in a chapter of a recent collaborative book. See René Lozi, Bo Zheng, Miguel Sanjuán, and Guanrong Chen, *Two more mathematicians and one physicist discuss their motivations and early career developments*, in *Dynamical Systems: Paths of Twelve Mathematicians and Physicists to Chaos and Its Applications*, edited by René Lozi, Safwan El Assad, and Mohammed-Salah Abdelouahab (CRC Press, 2025).

overview of this vibrant discipline to a broader international audience. After sharing the speech with colleagues from various institutions, I received additional encouragement to undertake this project, further reinforcing my conviction that the book needed to be written. I am especially grateful to Lakshmi for her sustained interest in my work, her unwavering support, and her invaluable encouragement throughout the process.

Though the book originated from my induction speech, I have substantially enriched and expanded its content. Most notably, I have included a dedicated chapter that addresses the crucial importance of teaching nonlinearity and complexity, with the sincere aim of making the basic concepts and ideas more accessible to a broader audience, particularly within the field of science. Numerous additions and updates have also been incorporated throughout the original Spanish text.

The book explores Nonlinear Dynamics, Chaos, and Complex Systems from multiple perspectives, including historical development, epistemological foundations, and the pioneering contributions of key scientific figures from the late 19th and early 20th centuries. I have taken particular interest in highlighting lesser-known ideas from several Nobel Laureates in Physics. It also delves into profound concepts such as emergence and indeterminism, highlighting their significance in physics and the life sciences.

Special attention is given to the interdisciplinary nature of this field, illustrating how its ideas permeate mathematics, physics, chemistry, biology, geology, and related scientific disciplines.

Precisely, these fields correspond to the three sections of the Royal Academy of Sciences of Spain: Mathematics, Physics and Chemistry, and Natural Sciences.

A significant portion of the book addresses broader themes related to complexity and interdisciplinarity, including the necessity of facilitating dialogue across disciplines and the potential of complexity as both an opportunity and a challenge in achieving true interdisciplinary integration.

Another topic of special interest, which I have chosen to include as a dedicated chapter, concerns the role of teaching complex systems thinking in academic curricula. It is of utmost importance to reflect on how complexity can be integrated into science education, especially in view of its significant implications for the future, particularly at the intersection of complexity, nonlinearity, and artificial intelligence.

The book concludes with two valuable appendices. The first appendix contains the official response speech delivered by Prof. Jesús María Sanz Serna, then President of the Royal Academy of Sciences of Spain, at my induction ceremony. The second appendix provides a list of selected publications from the research group I lead at Rey Juan Carlos University, highlighting interdisciplinary contributions to various aspects of Nonlinear Dynamics, Chaos, and Complex Systems.

It is my sincere hope that this book will not only offer readers a comprehensive and accessible overview of this dynamic field but will also stimulate curiosity and engagement among students, teachers and educators, researchers, scientists, and all those interested in scientific discovery. Ultimately, I hope it conveys the excitement, promise, and abundant opportunities that lie ahead in the fascinating world of Nonlinear Dynamics, Chaos, and Complex Systems.

Miguel A. F. Sanjuán

Villaviciosa de Odón, Madrid, June 2025

Preface

Hon. Mr. President of the Royal Academy of Sciences,

Excellencies Members of the Academy,

Ladies and Gentlemen,

I would like to express my satisfaction and deep gratitude to all the members of the Royal Academy of Exact, Physical and Natural Sciences of Spain for the distinction of having chosen me as a Full Academician. It is a great honor and a privilege to be part of this institution of prestige and academic excellence that has welcomed, over the last two centuries, since its foundation, an outstanding representation of Spanish science. At the same time, I assume with great responsibility the commitment to serve the institution.

My gratitude is very special to the Academicians Prof. Juan María Marcaide Osoro, Prof. Miguel Ángel Alario y Franco and Prof. Manuel Aguilar Benítez de Lugo for having been the ones who made the proposal for me to be part of the Royal Academy as a full member holding the number medal 60 newly created in the Physical and Chemical Sciences Section.

From the moment I was elected Corresponding Academician in 2015, I felt the desire to collaborate to the extent of my possibilities to the various initiatives of the Academy and to respond to the services that the institution has entrusted to me. Various opportunities have arisen to collaborate in areas such as communication and science dissemination, in addition to actively participating in different commissions, including the International Relations Commission and the Organizing Commission of the annual General Conference, a significant event for our Academy organization.

In this sense, my gratitude extends to the Supernumerary Academician and Former President of the Academy Prof. José Elguero Bertolini, to the President of the Academy, Prof. Jesús María Sanz Serna, as well as to the General Secretary of the Academy Prof. Ana María Crespo de las Casas, for the continuous support, affection and trust they have placed in me at all times. Furthermore, I especially thank Prof. Jesús María Sanz Serna for having agreed to make the response speech at my reception at the Academy.

In the scope of my scientific work, I want to make special mention of Prof. James A. Yorke of the University of Maryland, who is a Foreign Academician of the Mathematics Section of this Academy. Prof. Yorke is a true pioneer in chaos theory, being the one who coined the term *chaos* in the modern literature of Nonlinear Dynamics. His outstanding contributions earned him the prestigious *Japan Prize* in 2003 for his achievements in Complexity Science and Technology.

I recognize Prof. Yorke as my mentor and thank him enormously for the profound influence he has had on my academic career.

I would like to express my deep gratitude to all the people and institutions that have contributed to my training and have given me their support in my journey up to this point. A special thank you to my collaborators from the Nonlinear Dynamics, Chaos Theory and Complex Systems Group of the Rey Juan Carlos University, of which I have the honor of being creator and director, as well as to all my more than twenty doctoral students, from whom I have learned so much. Throughout my scientific career, I have had the privilege of collaborating closely with more than two hundred researchers of twenty-five different nationalities, to whom I also want to express my most sincere gratitude.

Needless to say, I sincerely appreciate the support that my parents and siblings have always given me, and especially my wife Céline and my daughters Alicia and Mónica.

In summary, this admission to the Royal Academy of Sciences of Spain represents a significant event in my career, and I feel grateful for the opportunity to contribute to the objectives and responsibilities of this Academy.

I conclude with a few words paraphrasing the great poet Dante Alighieri and the 1938 Nobel Prize in Physics, Enrico Fermi, who used a similar expression in analogous circumstances: **Nunc incipit vita nova, gaudeamus igitur**.

Miguel A.F. Sanjuán

Contents

Chapter 1

Introduction

Le savant n'étudie pas la nature parce que cela est utile; il l'étudie parce qu'il y prend plaisir et il y prend plaisir parce qu'elle est belle. Si la nature n'était pas belle, elle ne vaudrait pas la peine d'être connue, la vie ne vaudrait pas la peine d'être vécue[2].

— Henri Poincaré

1.1 Introduction

In this book, I will discuss a range of ideas and concepts, such as nonlinearity, chaos, complexity[3], and interdisciplinarity, that are deeply relevant not only in physics but also across the broader landscape of science. The field of Nonlinear Dynamics, Chaos Theory, and Complex Systems[4,5,6,7], an intrinsically interdisciplinary domain, has been

---

[2]"The scientist does not study nature because it is useful; he studies it because he takes pleasure in it, and he takes pleasure in it because it is beautiful. If nature were not beautiful, it would not be worth knowing, and life would not be worth living." Henri Poincaré. *Science et Méthode* (Ernest Flammarion, Paris, 1908), p. 20.

[3]Editorial. Complexity matters. Nature Physics 18, 843 (2022).

[4]James Gleick, *Chaos: Making a New Science* (Penguin Books, 2008).

[5]Ian Stewart, *Does God Play Dice? The New Mathematics of Chaos* (Penguin Books, 1997).

[6]Lenny Smith, *Chaos: A Brief Introduction* (Oxford University Press, 2007).

[7]Manuel de León and Miguel A. F. Sanjuán, *Mathematics and the Physics of Chaos* (La Catarata, Madrid, 2010). (in Spanish)

the primary focus of my entire scientific career, approached from the perspective of physics.

Dynamics may be regarded as one of the foundational sciences, given that it is intimately tied to the very emergence of the concept of motion. From the earliest attempts to understand how objects move, dynamics not only provided a systematic framework for describing motion but also laid the groundwork for the broader scientific ambition of predicting physical phenomena with precision. However, the notion of dynamics has evolved significantly over time. Traditionally, the term referred primarily to the motion of celestial bodies and rigid mechanical systems, grounded in classical mechanics. Today, when we speak of dynamics, we refer more broadly to any process that involves change over time. This broader definition means that dynamics now permeates all fields of science, including economics, chemical reactions, ecology, physiology, and neurodynamics, offering us a profoundly interdisciplinary perspective on nature. Moreover, the interactions among components of a system, along with feedback mechanisms, are key sources of nonlinearity and complexity. These features, combined with the idea of sensitive dependence on initial conditions, a hallmark of chaotic behavior, have led to a paradigm shift in our understanding of scientific systems, with far-reaching implications for how we model and interpret the natural world.

Throughout the text I intend to give a panoramic view of the discipline, presenting the fundamental concepts that have driven the development of the essential ideas, as well as the various sources that have contributed to its construction, as we understand it today.

I would like to highlight two fundamental aspects in this exhibition: interdisciplinarity and internationalization. Undoubtedly, science has been nonlinear since its inception: Newton's equations are nonlinear. However, the difficulties of dealing with nonlinear phenomena have frequently led to linear approximations when modeling certain problems that, logically, were easier to address. It is therefore not difficult to understand that most dynamical systems do not follow linear behavior.

To illustrate the construction of nonlinear dynamics, I like to think of it as the analogy of an extensive river fed by numerous tributaries. Among the most relevant problems are the three-body problem in celestial mechanics, turbulence in fluid dynamics, irreversibility and the foundations of statistical physics, the study of nonlinear oscillators, as well as the logistic map derived from population dynamics in biology. These have been the origins of this fascinating field of nonlinear dynamics, chaos and complex systems. It is equally astonishing to observe the numerous schools of mathematics and physics that have played an essential role in the development of this discipline, such as the French, Russian, Japanese and American schools. Understanding this historical perspective allows us to appreciate the breadth of the discipline itself and its multiple interdisciplinary applications in various fields of science.

Among the different situations that we can encounter in dynamics, we have equilibrium points, periodic and quasiperiodic motion and finally, chaotic behavior. Possibly one of the most profound insights into the nature of what is known as chaotic behavior is the idea of sensitive dependence on initial conditions. That is, the trajectories of a chaotic system move away from each other as time progresses when they start from very close initial points. This fact has very drastic consequences on the predictability of a system.

Perhaps one of the most recognized chaotic systems is the Lorenz system, which has been extensively studied and is possibly one of the best known in popular culture.

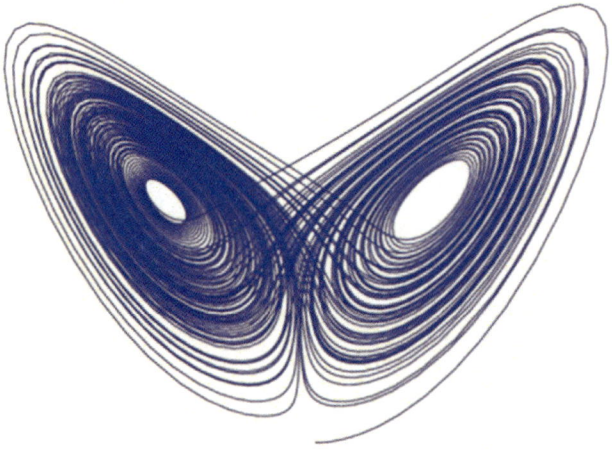

Fig. 1.1   The figure represents the chaotic Lorenz attractor.

Figure 1.1 presents a chaotic attractor of the Lorenz system in phase space. This system was introduced by the American meteorologist Edward Lorenz to investigate thermal convection in a fluid and, through numerical computer simulations, he was able to observe the property of sensitive dependence on initial conditions, which is distinctive of chaotic behavior.

Figure 1.2 shows the time evolution of two orbits (one red and one blue) whose initial conditions are very close. After a certain time,

approximately 24 time units, the corresponding orbits begin to separate, turning out to be very different in the long run, which reflects
very well the idea of sensitive dependence on initial conditions in the
chaotic Lorenz system.

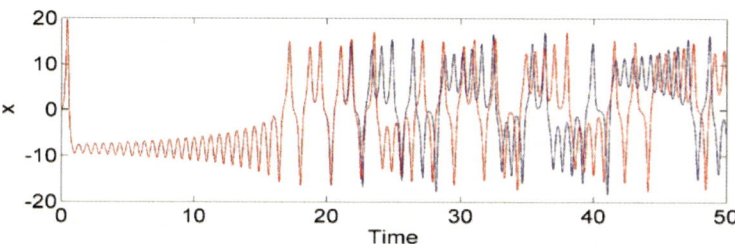

Fig. 1.2 Time evolution of two trajectories of the Lorenz system that start very close to each
other, illustrating the property of sensitive dependence on initial conditions.

1.2 Sensitivity Dependence on Initial Conditions

As soon as we think about this idea, we can have the intuition that
it is true in our lives. Which explains why it is not surprising that
the same idea has appeared in literature throughout history in multiple sources. In this sense, it is very pertinent to mention a famous
rhyme traditionally associated with Benjamin Franklin, although an-
tecedents of the same idea date back to the 15th century, and which
is known as *"For want of a nail"*, which offers an intuitive and poetic
image of the idea of sensitive dependence on initial conditions, which
is the hallmark of chaotic motion:

> "For want of a nail, the horseshoe was lost,
> For want of the horseshoe the horse was lost,
> For want of the horse, the rider was lost,
> For want of the rider, the message was lost,

For want of the message, the battle was lost,
For want of the battle, the kingdom was lost,
And all for the want of the horseshoe nail."

Particularly noteworthy is Henri Poincaré's discussion in Chapter 4 of his renowned 1908 work *Science et Méthode* (Science and Method), published in Paris and dedicated to the topic of chance, in which he offers a remarkable description of what is now known as sensitive dependence on initial conditions:

"Une cause très petite, qui nous échappe, détermine un effet considérable que nous ne pouvons pas ne pas voir, et alors nous disons que cet effet est dû au hasard. Si nous connaissions exactement les lois de la nature et la situation de l'univers à l'instant initial, nous pourrions prédire exactement la situation de ce même univers à un instant ultérieur. Mais, lors même que les lois naturelles n'auraient plus de secret pour nous, nous ne pourrons connaître la situation initiale qu'approximativement. Si cela nous permet de prévoir la situation ultérieure avec la même approximation, c'est tout ce qu'il nous faut, nous disons que le phénomène a été prévu, qu'il est régi par des lois; mais il n'en est pas toujours ainsi, il peut arriver que de petites différences dans les conditions initiales en engendrent de très grandes dans les phénomènes finaux; une petite erreur sur les premières produirait une erreur énorme sur les derniers. La prédiction devient impossible et nous avons le phénomène fortuit."[8]

---

[8]"A very small cause, which escapes us, determines a considerable effect that we cannot help but see, and then we say that this effect is due to chance. If we knew exactly the laws of nature and the situation of the universe at the initial moment, we could exactly predict the situation of that same universe at a later moment. But although natural laws no longer keep any secrets from us, we will only be able to know the initial situation approximately. If this allows us to predict the subsequent situation with the same approximation, that is all we need, we say that the phenomenon was foreseen, that it is governed by laws; but this is not always the case, it may happen that small differences in the initial conditions generate very large differences in the final phenomena; a small mistake at the beginning will produce a huge mistake at the end. Prediction becomes impossible and we have the fortuitous phenomenon." Henri Poincaré, *Science et Méthode* (Ernest Flammarion, Paris, 1908), pp 71-72.

Perhaps even more striking is a statement made by Charles Darwin in *The Origin of Species*[9]:

> "More individuals are born than can possibly survive. A grain in the balance will determine which individual shall live and which shall die, which variety or species shall increase in number, and which shall decrease, or finally become extinct."

which again illustrates the true notion of sensitive dependence on initial conditions.

1.3 About Interdisciplinarity. Relationships between the Different Sciences

In recent years, we have witnessed the growing phenomenon of super-specialization, which carries the risk of suggesting the notion that each field exists in isolation. In my view, however, the reality is quite the opposite: everything or nearly everything is interconnected. This recognition naturally leads us to the concept of interdisciplinarity and the intricate relationships that link the various branches of science. We could ask ourselves, at least rhetorically, about what the existing relationships are between physics, chemistry, biology, geology and mathematics, or how their interrelation benefits. As it is not easy to give a brief answer to some of these questions, a famous quote from the physicist Richard Feynman, the 1965 Nobel Prize in Physics, came to mind that illustrates this question:

> "A poet once said: *"The whole universe is in a glass of wine."* We will probably never know in what sense he meant it, since poets do

---

[9]Charles Darwin. *The Origin of Species*. 1859. Chapter XIV. http://www.talkorigins.org/faqs/origin.html.

not write to be understood. But it is true that, if we look at a glass of wine close enough, we see the entire Universe. There are the objects of physics: the liquid that moves and evaporates depending on the wind and time, the reflections in the glass and our imagination adds the atoms. The glass is a distillation of terrestrial rocks, and in its composition we see the secrets of the age of the universe and the evolution of the stars. What strange arrangement of chemical elements is there in wine? How did they come to be formed? There are ferments, enzymes, substrates and products. There in wine is the great generalization: all life is fermentation. [...] *Although, for convenience, our little minds divide this glass of wine, this universe into parts - physics, biology, geology, astronomy, psychology and so on - it is good to remember that nature does not know!*[10]So let's put everything back in its place, without ultimately forgetting what it is for. Let us allow ourselves one last pleasure: let us drink it and forget it all!"[11]

In my view, this clearly illustrates the idea that the academic disciplines we have developed over time constitute just one among many possible frameworks for understanding and exploring various aspects of the world around us. These established fields, though useful for organizing knowledge, often impose constraints on how we interpret phenomena by encouraging compartmentalized thinking. This tendency can lead us to overlook the deeper, more intricate truth that nearly everything is related to everything. Whether we are studying the natural world, human behavior, culture, or technology, it becomes increasingly evident that these categories frequently overlap

---

[10]I have put this phrase in italics to emphasize the idea of interdisciplinarity.

[11]R.P. Feynman, R.B. Leighton, and M. Sands. *The Feynman Lectures on Physics.* Vol. *I Mainly Mechanics, Radiation and Heat* (Addison-Wesley, Reading, Massachusetts, 1963), pp. 3-10.

and influence one another in complex, dynamic ways. Thus, adhering too rigidly to disciplinary boundaries may hinder our ability to grasp the full picture and appreciate the multifaceted nature of reality.

1.4 Physics and Complexity: Open Problems

In relation to the contents of this text, I consider it pertinent to evoke an event that took place in 2005 on the occasion of the celebration of the Seminar "100 years of Einstein's theories: present and future perspectives" that took place during June 1 and 2 at the CSIC Student Residence in Madrid on the occasion of the celebrations of the *2005 World Year of Physics*. Among the activities of the seminar, there was a Round Table dedicated to the analysis of "Open Problems", in which Prof. Alberto Galindo Tixaire, then President of the Royal Academy of Sciences of Spain, and Prof. Francisco José Yndurain Muñoz, then President of the Section of Physical and Chemical Sciences of the Royal Academy of Sciences of Spain, and the author of this book, acting as moderator Prof. Gerardo Delgado Barrio, who was President of the Spanish Royal Society of Physics. In that round table, new challenges and future perspectives were raised in the world of elementary particle physics, cosmology and the physics of complexity. My modest contribution was on "Physics and Complexity: Open Problems", where I described some key ideas related to the physics of complex systems and its future.

1.5 References and Mentions to RAC[12] Academicians

As I have already indicated, the contents of my speech are aimed at describing the discipline to which I have dedicated my scientific life. Although I am going to occupy a newly created medal, the name of the position on physics of complex systems being also new, I would like to briefly highlight those academicians who in one way or another have contributed or mentioned in their speeches to matters to which I will refer in my speech.

It has been more than 40 years since Prof. Alberto Galindo Tixaire's reception speech on *Nonlinearity in the natural sciences* took place[13]. Despite having read it several years ago, I cannot help but be amazed by its pioneering and premonitory nature on the role of nonlinearity and nonlinear phenomena in physics. Furthermore, he expresses his admiration for this field, highlighting its richness and the profusion of interdisciplinary connections with notable clairvoyance. In fact, in the conclusions he points out a whole prophecy: "Supplying with your imagination the aridity of my prose, I have no doubt that you will have already formed an idea of the varied panorama of non-linear phenomena. Its unifying nature of diverse disciplines ensures a brilliant future, in which all sciences from pure mathematics to advanced technology, side by side, mutually fertilizing each other, will reveal the wonderful richness of these processes and their unifying principles." I do not want to avoid mentioning

---

[12]RAC is the abbreviation for the Real Academia de Ciencias (Royal Academy of Sciences of Spain).

[13]Alberto Galindo Tixaire. *Nonlinearity in the natural sciences*. Reception speech. Royal Academy of Exact, Physical and Natural Sciences. June 11, 1980. https://rac.es/ficheros/doc/a758afa4503cb688.pdf (in Spanish).

other sentences like this one: "When the simplest of the laws of nature, the law of gravitation, leads to nonlinear equations that are difficult to analyze, it will surprise no one that the same or more happens in complex dynamical systems." And another in greater depth: "But also in other experimental sciences, such as chemistry, biology and medicine, a nonlinear description emerges with force, simplifying, into manageable mathematical schemes, enormously convoluted and sometimes poorly known dynamical behaviors."

Numerous references to nonlinear dynamics and its applications to biomedical sciences also appear in Prof. Pedro García Barreno's reception speech[14], as the following statement masterfully expresses: "Nonlinear dynamics offers, therefore, a language that makes it possible to describe different aspects of brain function, since the concepts of intermittency, alternating periods of periodic and chaotic behavior, transitivity or fluctuation, have fit in that context. It can be hypothesized that the set of general laws that govern the flows is universal, and that the brain is a possible location of abstract dynamical systems with spatial distribution. Let us remember that the EEG recording in a normal baseline situation is an example of chaotic behavior, while the recording corresponding to an epileptic seizure represents regular periodic behavior."

Likewise, in the speech of Prof. Amable Liñán Martínez[15] a reference is made to the phenomenon of chaos and the fractal properties

---

[14]Pedro García Barreno. *Of the Exact, the Physical, the Natural and Medicine.* Reception speech. Royal Academy of Exact, Physical and Natural Sciences. December 12, 1984. `https://rac.es/ficheros/doc/968629d78e167868.pdf` (in Spanish).

[15]Amable Liñán Martínez. *The role of fluid mechanics in combustion processes.* Reception speech. Royal Academy of Exact, Physical and Natural Sciences. January 23, 1991. `https://rac.es/ficheros/doc/c9dbd825cee2a0eb.pdf` (in Spanish).

of turbulent fluids in combustion reactions. In addition, he addresses the nonlinear theory of hydrodynamic stability, highlighting the chaotic behavior of simple dynamical systems, ergodic theory, strange attractors, and horizons of predictability in the description of fluids.

Similarly, it is full of references to nonlinear dynamics and chaos theory in chemistry, in general, and to chemical reactions in particular, the speech of Prof. Jesús Santamaria Antonio[16]. Not in vain the theoretical study of the chemical reaction cannot be explained without the context of the nonequilibrium statistical mechanics of irreversible processes that he mentions in his speech, as well as some of the pioneers of complexity in Physics. Likewise, the role that nonlinear dynamics plays in the context of energy transfer mechanisms is notable. The connection with nonlinear systems is clear: "Molecules are conservative Hamiltonian systems made up of coupled anharmonic oscillators, where frequencies vary with energy, and therefore, they are nonlinear systems." In this sense, it is not surprising that in his speech he makes continuous references to chaos theory, especially what is known as Hamiltonian chaos and KAM theory. In contrast to the traditional disciplinary divisions, the defense of interdisciplinarity is also notable, not only in the field of the dynamics of chemical reactions, but also in science in general: "The overcoming of these divisions, which occurred in the last decades of the 20th century,

---

[16] Jesus Santamaría Antonio. *Evolution of ideas around the chemical reaction.* Reception speech. Royal Academy of Exact, Physical and Natural Sciences. January 29, 2003. `https://rac.es/ficheros/doc/4e876e860a1bcf93.pdf` (in Spanish).

has led to the flourishing of interdisciplinary areas, not only within Chemistry but, in general, within the experimental sciences."

In the speech of Prof. Jesús María Sanz Serna[17] on *Geometric Integration*, numerical methods, symplectic methods in Hamiltonian systems, chaos and shadowing are mentioned, all of them being fundamental ideas in physics and nonlinear dynamics and chaos theory. In addition, a mention is made of the influential article by Sussman and Wisdom published in Science[18] where they tested the chaotic evolution of the solar system after arduous numerical simulations with dedicated computers, which also posed interesting challenges to mathematicians working on numerical methods.

In the speech on the theories of turbulence by Prof. Javier Jiménez Sendín[19], very familiar issues such as turbulence and chaos, the relevance of numerical methods as well as complexity in fluids are also described. Fluid mechanics and meteorology are precisely one of the paths that have given us the knowledge of chaos as a discipline that we have today. He logically dedicates a part of his speech to nonlinearity and dynamical systems, to the Lorenz system and to the Ruelle and Takens theory of turbulence, also advocating for the important role of numerical simulations.

---

[17] Jesús María Sanz Serna. *Geometric Integration*. Reception speech, Royal Academy of Exact, Physical and Natural Sciences, November 28, 2007. https://rac.es/ficheros/doc/d7b1648493aaf706.pdf (in Spanish).

[18] G.J. Sussman and J. Wisdom. Chaotic evolution of the solar system. Science 257, 52-62 (1992).

[19] Javier Jiménez Sendín. *Theories of Turbulence and Imagination in Physics*. Reception speech. Royal Academy of Exact, Physical and Natural Sciences. February 23, 2011. https://rac.es/ficheros/doc/01240.pdf (in Spanish).

Nonlinear diffusive processes are the object of Prof. Juan Luis Vázquez Suárez's speech[20], which are located in the field of partial differential equations. Although the methodology to address these processes is clearly different from that used in ordinary differential equations, the text is permeated with nonlinearity. Furthermore, its applications cover diverse fields such as physics, biology, engineering, statistical physics and complex dynamical systems, thus showing its relevance in multiple disciplines.

The speech of Prof. Pedro Miguel Echenique Landiríbar[21] devotes an entire chapter to glossing some of the ideas that I will talk about in my speech on complexity, emergence and reductionism, where he states a very clear idea about complexity that derives from the ideas of the 1977 Nobel Prize in Physics Phil Anderson in his article *More is Different* which I will talk about in my speech. "The physics of condensed matter teaches us that the passage from one level of complexity to a higher one is not possible simply through the direct application of the results of the previous level, but requires new ideas, new principles that are characteristic of the higher level." Other notable phrases from the speech are: "An essential task of the theoretical physics of the future will not be so much to write the ultimate equation as to understand emergent behavior in various ways, including, in due course, the emergence of life itself. We will

---

[20]Juan Luis Vázquez Suárez. Paths of Science. *From the Laplacian operator to nonlinear diffusive processes*. Reception speech. Royal Academy of Exact, Physical and Natural Sciences. March 26, 2014. `https://rac.es/ficheros/doc/01233.pdf` (in Spanish).

[21]Pedro Miguel Etxenike Landiribar. *Dynamics of ions and electrons in solids and surfaces and small touches on science*. Reception speech. Royal Academy of Exact, Physical and Natural Sciences. May 31, 2017. `https://rac.es/ficheros/doc/01235.pdf` (in Spanish).

continue to preserve the precious values of reductionism, but delving deeper and deeper into the emergence that arises from complexities of all kinds." [...] "The two visions, reductionist and emergentist, complement each other and are not enemies. "We need reductionism with emergence and emergence with reductionism."

An idea that Prof. Luis Manuel Liz Marzán[22] expresses in his speech has powerfully caught my attention, which I want to review because it is very pertinent in relation to the content of my speech: "Terms such as fractals, critical fluctuations, microemulsions, chaos theory or synergetics, were completely foreign to us and immediately attracted my attention when selecting the topic of work for what was then called "Tesina de Licenciatura (sort of Bachelor's thesis)". Were we being guided towards the concept of interdisciplinarity? Probably, although I am not aware of having used that term at that time."

Given the contents of my speech and the intrinsic interdisciplinary nature of nonlinear dynamics, chaos theory and complex systems, it is of special interest to mention some phrases from the opening address of the 2014-2015 academic year by Prof. Carlos Belmonte Martínez[23]: "However, to confirm these speculations it will be necessary to individualize the components of the unified activity of the

---

[22]Luis Manuel Liz Marzán. *Plasmonic nanomaterials and nanomedicine. Multidisciplinary science in the 21st century.* Reception speech. Royal Academy of Exact, Physical and Natural Sciences. October 27, 2021. https://rac.es/ficheros/doc/a8fb2424cceff436.pdf (in Spanish).

[23]Carlos Belmonte Martínez. *Unraveling the brain. The paths of neuroscience.* Opening address of the 2014-2015 academic year. Royal Academy of Exact, Physical and Natural Sciences. October 16, 2014. https://rac.es/ficheros/doc/01218.pdf (in Spanish).

brain and model the computational capabilities of the interactive neural networks that form it. In 1943, Warren McCulloch and Walter Pitts mathematically designed the first artificial neuron and postulated that thought could be reduced to the basic concepts of binary logic.". [...] "Modern neurobiology has approached the brain, both from a reductionist perspective that tries to understand the whole by breaking it down into parts, and with holistic approaches, which try to explain its functioning in an integrated way. Although reductionists are criticized for falling into a simplifying determinism and holists are reproached for proposing, without a mechanistic basis, speculative theoretical constructs about cognitive functions, both approaches are valid and complementary to scientifically understand the brain." Finally, he points out in a visionary way that "It is not unreasonable to imagine that, in a few years, the study of brain computing mechanisms will pass from biologists to physicists, mathematicians and computer scientists, who will have to lead the attempt to model cognitive aspects of the brain and clarify whether our thinking also has non-computable components, as artificial intelligence researchers still discuss."

I would like to end this part now to tell a personal anecdote that I experienced in 1990 when on September 10 I decided to write to Prof. Ángel Martín Municio, President of the Royal Academy of Sciences (RAC) and Academician of the Royal Spanish Academy (RAE) (Spanish Language Academy), suggesting including the word *Chaos* in the Dictionary of the Royal Spanish Academy. My proposal was: "Chaos: (3) mat. (fis.) Stochastic behavior in a deterministic system." The Academician of the RAE, Prof. Rafael Alvarado, responded to me after some months, informing me that it would be considered by the Technical Vocabulary Commission of the RAE and that in any case it would go to the archives of the RAE as the original authority. What would be my surprise to see that in the 22nd edition of the Dictionary of the Royal Spanish Academy the third meaning of the word was included as: "Chaos: 3. m. Phys. and Mat. Apparently erratic and unpredictable behavior of some deterministic dynamical systems with great sensitivity to initial conditions."

Chapter 2

In the Beginning

Never in the annals of science and engineering has there been a phenomenon so ubiquitous, a paradigm so universal, or a discipline so multidisciplinary as that of chaos. Yet chaos represents only the tip of an awesome iceberg, for beneath it lies a much finer structure of immense complexity, a geometric labyrinth of endless convolutions, and a surreal landscape of enchanting beauty. The bedrock which anchors these local and global bifurcation terrains is the omnipresent nonlinearity that was once wantonly linearized by the engineers and applied scientists of yore, thereby forfeiting their only chance to grapple with reality[24].

— Leon Chua

2.1 Introduction

Exploring the origins of any subject is rarely a straightforward task, and the beginnings of a discipline are certainly no exception. As outlined in the introduction, my aim is to provide a panoramic and historical perspective, one that allows us to appreciate not only the depth of the field but also the diversity of its applications across the sciences.

---

[24]Leon Chua. International Journal of Bifurcation and Chaos 1(1), 1-2 (1991).

2.2 Instabilities in Fluids and Singular Solutions

Claude Navier (1785-1836), Barré de Saint-Venant (1797-1886), and Joseph Boussinesq (1842-1929) were distinguished engineers and mathematicians, each recognized for their significant contributions to their respective fields. Claude Navier, a French engineer and physicist, made pivotal advances in the theory of elasticity and fluid mechanics. The Navier-Stokes equations, which describe the motion of fluids, are named in his honor, shared with the English physicist George Stokes[25]. Barré de Saint-Venant, also a French engineer and mathematician, is renowned for his work in fluid mechanics and elasticity theory, notably formulating the Saint-Venant equations for water flow in open channels. Joseph Boussinesq, a French mathematician and physicist, is celebrated for his contributions to fluid theory and geophysics, particularly the Boussinesq equations, which are used to model the behavior of waves in fluids. All three have left a lasting legacy in their respective fields, and their contributions remain fundamental in physics and engineering. In their studies on the mathematical modeling of fluids, they were fascinated by the investigation of the so-called singular solutions of some differential equations, where infinitely small differences would produce enormously different effects. Karl Pearson, one of the founders of mathematical statistics wrote "I have a letter from Clerk Maxwell in which he says

---

[25] Mark McCartney, Andrew Whitaker and Alastair Wood (eds.) *George Gabriel Stokes. Life, Science and Faith.* (Oxford: Oxford University Press, 2019).

that the work of Saint-Venant and Boussinesq on singular solutions is marking a milestone. . . "[26]

2.3 Determinism, Singularities and Instabilities

Throughout the 19th century, certain limitations appeared around the myth of determinism. On the one hand, it is essential to have a complete understanding of the initial conditions of the problem. On the other hand, notable difficulties arose in resolving the dynamics of a physical system made up of a large number of particles. The latter led to the introduction of concepts related to probability theory in the study of the physical laws of systems formed by many particles, such as gases, liquids and solids, giving rise to the birth of statistical mechanics. Founding fathers of the discipline include Ludwig Boltzmann (1844-1906), Josiah Willard Gibbs (1839-1903), and James Clerk Maxwell (1831-1879), whose visionary work laid the groundwork for modern theories of complexity, thermodynamics, and beyond.

2.4 About Maxwell and Chaos

The Scottish physicist James Clerk Maxwell (1831-1879) is known primarily for having unified the laws of electricity and magnetism into a single elegant framework, what we now call Maxwell's equations. However, his contributions to physics have been among the

---

[26]Ian Hacking. *The Taming of Chance* (Cambridge: Cambridge University Press, 1990).

most prolific in the history of science. Among his great scientific work, it is important to mention that he is considered the father of automatics and statistical mechanics. However, the role he played in the development of modern chaos theory is largely unknown.

From one of his writings: *Does the progress of physical science tend to give any advantage to the view of necessity (or determinism) over that of the contingency of events and free will?*[27] delivered as a lecture in Cambridge on February 11, 1873, the following extracts show the extent to which James Clerk Maxwell was aware of what we today call *sensitive dependence on initial conditions*, which is the hallmark of chaos in a physical system. His insights, far ahead of their time, demonstrate an early awareness of the subtleties that would later shape modern chaos theory.

"Much light may be thrown on some of these questions by the consideration of stability and instability. When the state of things is such that an infinitely small variation of the present state will alter only by an infinitely small quantity the state at some future time, the condition of the system, whether at rest or in motion, is said to be stable; but when an infinitely small variation in the present state may bring about a finite difference in the state of the system in a finite time, the condition of the system is said to be unstable. It is manifest that the existence of unstable conditions renders impossible the prediction of future events, if our knowledge of the present state is only approximate, and not accurate."

---

[27] James Clerk Maxwell, 'Does the Progress of Physical Science tend to give any Advantage to the Opinion of Necessity (or Determinism) over that of the Contingency of Events and the Freedom of the Will?', in The Life of James Clerk Maxwell, with selections from his correspondence and occasional writings, by Lewis Campbell and William Garnett (MacMillan and Co., London, 1884), pp. 357-366.

This same opinion has been defended by Brian Hunt and James Yorke in the article entitled *Maxwell on Chaos*[28], where it is quoted verbatim: "It is our view that Maxwell was the first person to understand chaos, that is, he recognized the existence and importance of systems with "sensitive dependence to initial data." We will let Maxwell explain his understanding in his own words."

Maxwell concludes his 1873 essay by encouraging his scientific colleagues not to limit themselves to the study of nonchaotic models of physical systems: "...but it is to be expected that in phenomena of higher complexity there will be a far greater number of singularities, near which the axiom about like causes producing like effects ceases to be true.." [...] "If, therefore, those cultivators of physical science from whom the intelligent public deduce their conception of the physicist, and whose style is recognised as marking with a scientific stamp the doctrines they promulgate, are led in pursuit of the arcana of science to the study of the singularities and instabilities, rather than the continuities and stabilities of things, the promotion of natural knowledge may tend to remove that prejudice in favour of determinism which seems to arise from assuming that the physical science of the future is a mere magnified image of that of the past."[29]

A very important tool in nonlinear dynamics is the geometric notion of phase space. The notion of phase space[30] is attributed to the American physicist Josiah Willard Gibbs (1839-1903), who was

---

[28]Brian R. Hunt and James A. Yorke. Maxwell on Chaos. Nonlinear Science Today 3(1), 1-4 (1993).

[29]Lewis Campbell and William Garnett. *The Life of James Clerk Maxwell, with selections from his correspondence and occasional writings* (MacMillan and Co., London, 1884), pp. 364 and 366.

[30]D.D. Nolte. The tangled tale of phase space. Physics Today 63 (4), 33–38 (2010).

one of the pioneers of kinetic theory and is also considered one of the founding fathers of statistical mechanics, a term he also coined. The concept of phase space plays a crucial role in nonlinear dynamics, from the analysis of which we can obtain much information about a given dynamical system. The study of the phase space of a dynamical system allows us to obtain complex fractal structures whose physical consequences are reflected in the uncertainty associated with the prediction of the future state of the system.

Statistical mechanics is a fundamental branch of theoretical physics that seeks to describe the macroscopic properties of large systems of particles in terms of their averaged microscopic behavior. It combines the basic laws governing the dynamics of particles with statistical principles, particularly those related to the law of large numbers. The discovery of deterministic chaos has led some physicists to revisit the foundations of statistical mechanics from a new perspective. This is because deterministic chaos suggests that even systems with very few degrees of freedom, far from the thermodynamic limit, can exhibit complex behavior that necessitates statistical approaches. Over the past century, considerable efforts have been devoted to clarifying the dynamical origins of irreversibility. Yet, despite this progress, there remains no general consensus on the essential ingredients required to firmly establish the theoretical foundations of statistical mechanics.

The problem of irreversibility was one of the main concerns of one of the "founding fathers" of statistical mechanics (Fig. 2.1), the Viennese physicist Ludwig Boltzmann (1844-1906).

Fig. 2.1 The founding fathers of Statistical Mechanics: Ludwig Boltzmann (1844–1906), James Clerk Maxwell (1831–1879), and Josiah W. Gibbs (1839–1903).

The objection raised by Josef Loschmidt (1821-1895) to Boltzmann's program, consisting of deriving the laws of thermodynamics directly from mechanical behavior, revealed the paradox of a situation in which, while the laws of mechanics are reversible under time inversion, the thermodynamic behavior of systems is fundamentally irreversible. Certainly, in this century much progress has been made in the attempt to clarify the dynamical origin of the kinetic equations, although the problem remains open to a certain extent. Following Boltzmann, the first attempts to substantiate classical statistical mechanics were based on the supposed validity of the ergodic hypothesis, which, after considerable theoretical efforts, led to a real dead end. Following the work of Maxwell and Boltzmann, Gibbs introduced the concept of *mixing* associated with a system that uses the simile of an oil drop in an immiscible fluid, a small region in the phase space that simulates the oil drop, the dynamical evolution would help fill the entire phase space.

This idea implies that for a given dynamical system, two sufficiently close points would separate exponentially after a certain period of time, linked to the notion of the sensitive dependence on initial conditions and chaotic dynamics. One of the quantitative methods to determine when a dynamical system is chaotic is through the so-called *Lyapunov exponents*[31], which derive from the work of the Russian mathematician Alexander M. Lyapunov (1857-1918) on the stability of motion with an enormous influence on Physics. His ideas were published in the famous article *The general problem of the stability of motion*[32]. When a dynamical system has a positive Lyapunov exponent, it is considered chaotic, indicating that long-term prediction of its evolution becomes fundamentally impossible due to sensitivity to initial conditions.

2.5   Poincaré, the Three-Body Problem and the Birth of Chaos

To understand the three-body problem[33], we have to go back to the beginnings of modern science with the works of Isaac Newton on the gravitational field and the universal law of gravitation. The so-called two-body problem basically consists of analyzing the motion of a system formed by two bodies that attract each other under the action of gravitational forces. Newton solves the problem by reducing the motion of the two bodies to the motion of each of them

---

[31] Arkady Pikovsky and Antonio Politi. *Lyapunov exponents: a tool to explore complex dynamics* (Cambridge University Press, Cambridge, 2016).

[32] А. М. Ляпуновъ, "Къ вопросу объ устойчивости движенія", Сообщ. Харьков. матем. общ. Вторая сер., 3 (1893), 265–272. A. Liapunoff. General problem of motion stability. Annales de la faculté des sciences de Toulouse 2e series, tome 9 (1907), p. 203-474. AM Lyapunov. The general problem of the stability of motion. Int. J. Control. 55(3), 531-773 (1992).

[33] M. Valtonen, J. Anosova, K. Kholshevnikov, A. Mylläri, V. Orlov, K. Tanikawa. *The Three-body Problem from Pythagoras to Hawking* (Springer, Cham, 2016).

around the so-called center of mass, which is a point whose mass is the total mass of the system.

Later, an attempt was made to solve the three-body problem, which can be formulated in a simple way: suppose that 3 bodies of arbitrary masses m_1, m_2 and m_3 are mutually attracted by Newton's law of gravitation. Assuming that they can move freely in a three-dimensional space and with arbitrary initial conditions, determine the evolution of the movement. Despite the simplicity of its formulation, attempts to solve it have caused real headaches for many scientists. Among them we can highlight Isaac Newton (1642-1727), Alexis Clairaut (1713-1765), Leonhard Euler (1707-1783), Pierre-Simon Laplace (1749-1827), Joseph-Louis Lagrange (1736-1813), Carl Jacobi (1804-1851), George Hill (1838-1914) and Henri Poincaré (1854-1912).

It is precisely the latter who wrote a famous memoir in 1889 on the three-body problem and the equations of dynamics[34], after winning the prize in the competition on the stability of the solar system convened by King Oscar II of Sweden and Norway on the occasion of his 60th anniversary. This contest[35] had been proposed by the Swedish mathematician Gösta Mittag-Leffler[36], who had received from his teacher, the German mathematician Karl Weiertrass, the idea that the contestants write an original work facing one of four questions. One of Weierstrass's four questions had to do with

---

[34]Henri Poincaré. Sur le problème des trois corps et les équations de la dynamique. Acta Mathematica 13, 1-270 (1890).

[35]M. Mikael Rågstedt. From Order to Chaos: The Prize Competition in Honor of King Oscar II. http://www.mittag-leffler.se/library/prize-competition.

[36]Arild Stubhaug. Gösta Mittag-Leffler. *A Man of Conviction* (Springer, Berlin, 2010).

Celestial Mechanics. The question was born from a suggestion formulated by the mathematician Peter Gustav Lejeune Dirichlet of the University of Göttingen, who in 1858 had told his student Leopold Kronecker that he had discovered a new method for solving certain differential equations and noted that by applying them to the equations of celestial mechanics he could demonstrate with all rigor that the solar system was stable. The committee that evaluated this contest was made up of three mathematicians: the German Karl Weiertrass, the Frenchman Charles Hermite and the Swede Gösta Mittag-Leffler. Poincaré's motto was: *"Nunquam praescriptos transibunt sidera ends."* [37]

Later, in 1892, Poincaré published his great work *Les Méthodes Nouvelles de la Mécanique Céleste*[38] (Fig. 2.2) in three volumes where numerous new concepts appear that have given rise to the development of the theory of dynamical systems, as mathematicians usually call it or nonlinear dynamics, a term most used by physicists, as well as by other mathematical disciplines such as topology, or *analysis situs* as it was called at the time.

This is one of the reasons why Poincaré is considered one of the fathers of chaos theory, since many fundamental ideas of the theory are contained in this book.

Despite the difficulty of the general three-body problem, there is a case called restricted, circular and plane, which is the one that has been studied by many of the scientists I referred to above.

---

[37]Nothing exceeds the limits of the stars.
[38]Henri Poincaré. *Les Méthodes Nouvelles de la Mécanique Céleste* (Gauthiers-Villars et Fils, Paris, 1892, 1893, 1899).

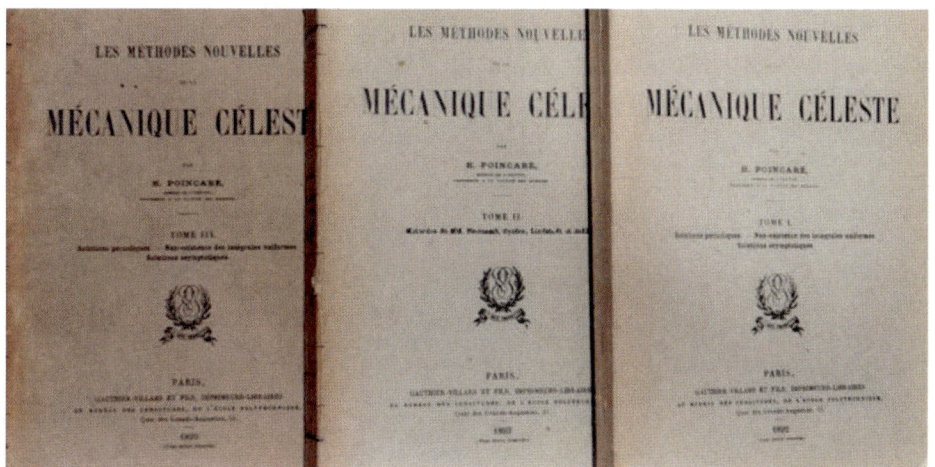

Fig. 2.2   The New Methods of Celestial Mechanics in three volumes were published by Henri Poincaré in 1892, 1893, and 1899.

Fundamentally, it is assumed that the system is not made up of any three masses, but that one of them is assumed to be much larger than the others and the third of them is of negligible mass compared to the rest. The analogy certainly comes from considering systems like the Sun, Earth, and Moon, or the Earth, Moon, and a satellite, where the approximation of moving in a plane is also correct. Under these circumstances and with an appropriate reference system, the equations of motion can be found without difficulty, from which a potential is derived that gives us an idea of the equilibrium positions in which a third body can be found. These are the five equilibrium positions that Lagrange found, which is why they are currently known as Lagrange points (Fig. 2.3).

Knowledge of the Lagrange points is very useful. In fact, at point L1 is the *Solar and Heliospheric Observatory (SOHO)*, which is a space probe to study the Sun. On January 6, 2024, *Aditya-L1*

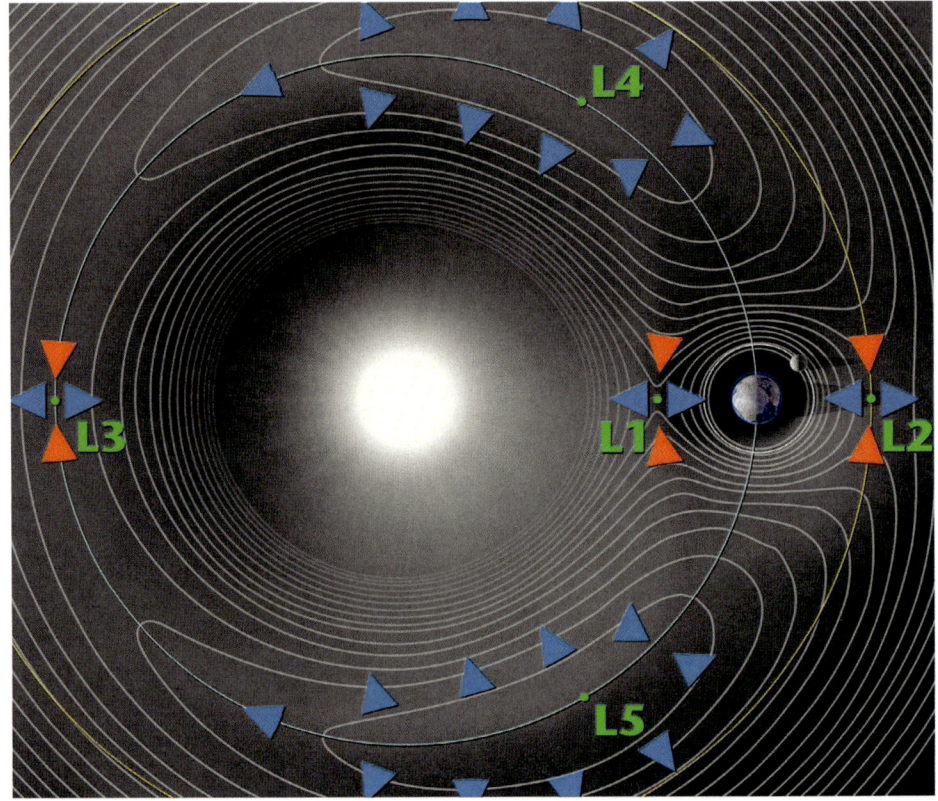

Fig. 2.3   The figure shows the equipotential curves of the restricted three-body problem, in this case, Sun-Earth-Moon, illustrating the five Lagrange points.

entered the halo orbit around the Lagrange point L1. This is a solar observation mission of the Indian Space Research Organization (ISRO) which will study the solar atmosphere, solar magnetic storms and their impact on the environment around Earth. At the Lagrange point L2, the *Wilkinson Microwave Anisotropy Probe (WMAP)* was positioned to study the microwave background radiation of the universe, managing to keep it in place with minimal fuel consumption, always keeping its sensors pointed away from the Earth and the Sun. In 2021, the *James Webb Space Telescope (JWST)*, which is a space observatory that will study the sky in infrared frequency and orbits around the Lagrange point L2. In 2022, the *Euclid mission* was launched by ESA, which also orbits around the same L2 point.

Fig. 2.4   The French mathematician and physicist Henri Poincaré (1854–1912).

As noted above, Poincaré (Fig. 2.4) did not address the three-body problem in a general way, but instead focused on studying what is known as the *restricted three-body problem*, which is a particular case in which one of the masses is considered to be very small compared to the others. In this study he found what he called doubly asymptotic or homoclinic orbits, which are characterized by having a homoclinic point in phase space. The presence of one of these points has very serious implications on the dynamic complexity of the system. After studying the problem, Poincaré wrote: *''On sera frappé de la complexité de cette figure, que je ne cherche même pas à tracer. Rien n'est plus propre à nous donner une idée de la complication du problème des trois corps et en général de tous les problèmes de Dynamique."* [39]

And in trying to solve this problem, he created a geometric method through which he saw that this problem had a very complex dynamics that is basically what we now call deterministic chaos. Figure 2.5 shows the invariant manifolds of a fixed point whose intersection gives rise to the appearance of a homoclinic point, generating what is known as the chaotic tangle.

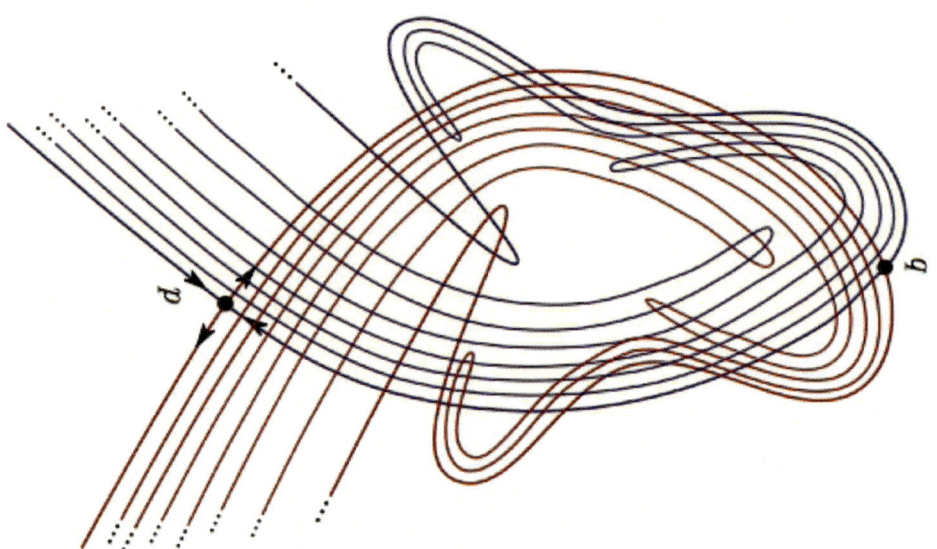

Fig. 2.5   Representation of the so-called chaotic tangle, showing a homoclinic point and the figure Poincaré referred to.

Poincaré's influence on the development of Hamiltonian systems is very significant. In this sense, it is interesting to note that his legacy was continued by the American mathematician George David

---

[39]"We will be surprised by the complexity of this figure, which I do not even attempt to draw. Nothing is more appropriate to give us an idea of the complication of the three-body problem and, in general, of all the problems of Dynamics." Henri Poincaré. *Les Méthodes nouvelles de la Mécanique Céleste*, Tome 3, 1899.

Birkhoff (1884-1944), who is credited with coining the term dynamical systems[40]. Birkhoff, in turn, exerted considerable influence on Edward Lorenz, who would rediscover sensitive dependence on initial conditions in the mid-20th century.

Within this current of thought, and in the American context, it is necessary to mention the mathematician Steven Smale (Fig. 2.6), who received the Fields medal in 1966 for his great contributions to the theory of dynamical systems. Precisely, we owe to him the concept of *Smale horseshoe*, which constituted an important step in understanding the relationship between the existence of a homoclinic point and the notion of deterministic chaos, through the simple idea of symbolic dynamics by using the so-called *Bernoulli shift map*.

Fig. 2.6   Steve Smale, Fields Medal 1966.

---

[40]G.D. Birkhoff. *Dynamical Systems* (American Mathematical Society, Providence, RI, 1927).

Another key school within this same tradition is that of Andrei N. Kolmogorov (1903-1987). All of them developed new methods and made notable contributions to the construction of nonlinear dynamics. In 1954, at the International Mathematics Congress held in Amsterdam, Kolmogorov stated a theorem for Hamiltonian systems that was later proven by his student Vladimir I. Arnold (1937-2010) and by the German Jürgen Moser (1928- 1999), which has turned out to be of considerable importance. This theorem is currently known as the KAM (Kolmogorov-Arnold-Moser) theorem[41] and it has to do with the problem of the stability of invariant tori in integrable systems of Hamiltonian mechanics under the action of small perturbations. This work, in fact, links naturally with Poincaré's pioneering work on celestial mechanics, since he had brought to light the idea of the complexity of orbits in the three-body problem, and the KAM theorem can be considered a culmination of these ideas. As we have already seen, the stability of the solar system is a problem of special importance in celestial mechanics and the KAM theorem shows that under certain conditions these orbits remain confined to certain regions.

In the excellent review article by Phil Holmes[42] the story connecting Poincaré, with celestial mechanics, the theory of dynamical systems, and chaos is described. As early as 1890, Henri Poincaré recognized that the presence of homoclinic points in phase space,

---

[41]H. Scott Dumas. *The KAM Story. A Friendly Introduction to the Content, History, and Significance of Classical Kolmogorov – Arnold – Moser Theory* (World Scientific, Singapore, 2014).

[42]P. Holmes. Poincaré, celestial mechanics, dynamical-systems theory and "chaos". Physics Reports 193(3), 137-163 (1990).

arising from the transversal intersection of the stable and unstable manifolds of a saddle point, constituted a fundamental mechanism for the onset of chaotic behavior. He understood that the breakdown of separatrices and the resulting complex structure of the phase space implied sensitive dependence on initial conditions and a loss of predictability in the system's long-term evolution. Melnikov's method[43] allowed proving the existence of transversal homoclinic points in Poincaré maps that arise from specific examples of perturbed differential equations, which contributed to providing a perturbative analytical method to determine conditions for the appearance of chaos in nonlinear dynamical systems. The transition from a regular system to a chaotic system is quite abrupt. In Hamiltonian systems this transition is due to a phenomenon known as *resonance overlap criterion*[44] mainly due to the work of the Greek astronomer George Contopoulos. If a perturbation goes beyond a certain critical value, the system can become very chaotic as in the case of the Hénon-Heiles Hamiltonian[45]. Finally, the method became popular with the work of the Russian physicist Boris Chirikov and came to be called the Chirikov criterion[46]. The Hénon-Heiles Hamiltonian is a model of galactic dynamics and constitutes a paradigm in nonlinear dynamics in two-dimensional and time-independent Hamiltonian systems,

[43]V.K. Melnikov. On the stability of the center for time periodic perturbations. Trans. Moscow Math. Soc. 12, 1-57 (1963).

[44]G. Contopoulos. Resonance phenomena and the non applicability of the "third" integral. Bulletin Astronomique 2, 223-241 (1967).

[45]M. Hénon and C. Heiles. The applicability of the third integral of motion: some numerical experiments. Astronomical J. 69, 73-79 (1964).

[46]B.V. Chirikov. A universal instability of many-dimensional oscillator systems. Phys. Rep. 52, 263 (1979).

which has been widely used to study Hamiltonian numerical chaos, as well as fractal structures in phase space[47].

2.6 Fractional Dimensions, Fractals and Chaos

The development of fractal geometry initiated by the physicist and mathematician Benoit Mandelbrot[48](1924-2010) (Fig. 2.7), who had been a student of the French mathematician Gaston Julia, has played a fundamental role in the understanding and analysis of the complex behavior of nonlinear dynamical systems. In any case, it is important not to forget the role played in many aspects of the development of nonlinear dynamics by the German mathematician Georg Cantor (1845-1918), particularly with respect to the Cantor middle-third set and its constant appearance in many dynamical problems.

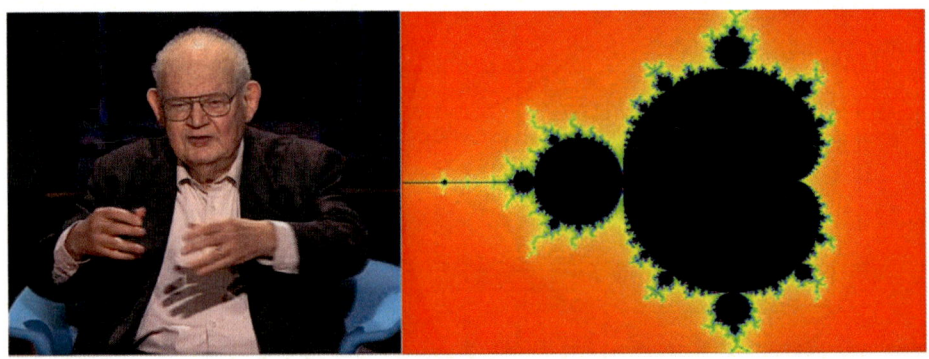

Fig. 2.7   Benoit Mandelbrot (1924–2010) and the famous set that bears his name.

There are many complex geometric shapes in nature, such as coastlines, river beds, biological shapes, and even the complex curves

---

[47]J. Aguirre, J.C. Vallejo, M.A.F. Sanjuán. Wada basins and chaotic invariant sets in the Hénon-Heiles system. Phys. Rev. E 64 (6), 066208 (2001).
[48]Benoit B. Mandelbrot. *The Fractalist: Memoir of a Scientific Maverick*. Vintage. (2014).

of financial markets. A common characteristic in all of them is self-similarity. This is the property that when a part is enlarged in this way, the same type of structure appears. To characterize objects with this universal property, the use of fractional dimensions is necessary, which led Benoit Mandelbrot to call these objects *fractals*. The enormous task of compiling the work carried out by mathematicians before him, such as the Frenchman Gaston Julia, the Swede Helge von Koch, the Polish Wacław Sierpiński, as well as the works on dimensions by the German Felix Hausdorff and the Russian Abram S. Besikovich had a notable influence creating the field of fractal geometry.

The Koch curve was devised by Swedish mathematician Helge von Koch (1870-1924), and is constructed iteratively by dividing a line segment into three equal parts and replacing the middle segment with two larger segments that form an equilateral triangle. This process is repeated in each new segment, generating a self-similar fractal. The Sierpinski fractal is due to the Polish mathematician Wacław Sierpiński (1882-1969) and is formed by the successive elimination of smaller triangles inside an initial equilateral triangle. In each iteration, triangles are removed, leaving a recursive pattern. The process is repeated, creating a geometric design with self-similarity at different scales.

The notion of dimension is fundamental when measuring geometric objects. There are several ways to define the concept of dimension, but it is clear that a point has dimension zero, a straight line has dimension one, a plane has dimension two, and a cube has dimension three. However, and as strange as it may seem, there are geometric objects whose dimensions are not an integer, turning out to be a fractional value. This is a simple notion of what is meant by a fractal dimension or a Hausdorff dimension, so that the Cantor set, mentioned above, has a dimension of $log2/log3 \approx 0.63$, the Koch curve has a dimension of $log3/log4 \approx 1.26$ and the Sierpinski set has a dimension of $log3/log2 \approx 1.585$ (Fig. 2.8). All of them are self-similar fractal sets, since they are obtained through an iterative rule so that the basic structure is repeated at all scales.

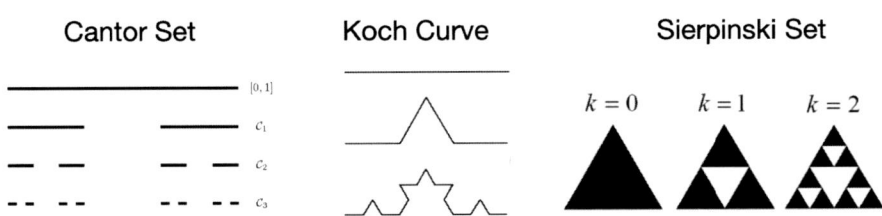

Fig. 2.8　The figure shows the Cantor set, the Koch curve and the Sierpinski set.

Although in principle fractal geometry and nonlinear dynamics are two disciplines that apparently have nothing to do with each other, however, as noted above, chaos and fractals[49] are closely linked. One of the main ideas is due to the fact that associated with the notion of chaos there is that of the chaotic attractor, which constitutes a geometric object of fractal nature that lives in phase space, so it is impossible to talk about chaos without talking about fractals and vice versa.

[49]H.-O. Peitgen, H. Jürgens, and D. Saupe. *Chaos and Fractals. New Frontiers of Science* (Springer, New York, 2004), Second edition.

2.7   Complexity in Fluid Motion

The phenomenon of turbulence in fluid motion is one of the most spectacular cases of chaotic behavior. The American mathematician Norbert Wiener[50] called it "the wave problem," which he had discovered at Cambridge in the writings of Geoffrey I. Taylor[51] that would later inspire studies on chaos[52]. Although the Navier-Stokes equations, which are the fundamental equations of fluid motion, have been known since the end of the 19th century, it must be remembered that the form of their solutions in the turbulent regime is not yet known. The Clay Mathematics Institute proposed it in 2000 as one of the millennium problems. In 1963, Massachusetts Institute of Technology (MIT) meteorologist Edward N. Lorenz[53](Fig. 2.9) developed a model of three ordinary differential equations to describe the movement of a fluid under the action of a thermal gradient.

When finding numerical solutions with the help of a computer, he once again encountered the phenomenon of sensitive dependence on initial conditions. That is, the system was inherently unpredictable, such that small variations in the determination of the initial conditions led to drastically different solutions. At the time, very few gave importance to this fact, perhaps because the results of Lorenz's work were published with a somewhat cryptic title, *Deterministic*

---

[50]Norbert Wiener. *I Am a Mathematician* (MIT Press, 1964).
[51]G.K. Batchelor. *The life and legacy of G. I. Taylor* (Cambridge University Press, 1996).
[52]Flo Conway and Jim Siegelman. *Dark hero of the information age. In search of Norbert Wiener the father of cybernetics* (Basic Books, New York, 2005).
[53]T.N. Palmer. Edward Norton Lorenz. May 23, 1917 — April 16, 2008. Biogr. Mems Fell. R. Soc. 55, 139–155 (2009).

Fig. 2.9   Edward N. Lorenz (1917-2008).

Nonperiodic Flow[54] in a meteorology journal and went unnoticed by many scientists.

The theory of the Russian physicist Lev D. Landau, and the German mathematician Eberhard Hopf, which proposed the existence of an infinite set of incommensurable frequencies to explain turbulence, was surpassed in the 1970s by the theoretical contributions of David Ruelle and Floris Takens, who introduced the fundamental concept of the *strange attractor* in 1971[55]. It is an attractive geometric

---

[54]E.N. Lorenz (1963). Deterministic Nonperiodic Flow. Journal of Atmospheric Sciences 20, 130-148 (1963).
[55]D. Ruelle and F. Takens. On the Nature of Turbulence. Commun. Math. Phys. 20, 167-192 (1971).

object, distinct from previously known cases such as periodic fixed points, quasi-periodic orbits, or limit cycles, hence the term "strange," and it also possesses a fractal (non-integer) dimension.

Very shortly after, the German physicist Hermann Haken wrote an article[56], where he establishes an analogy between the equations of laser dynamics and the Lorenz model, being one of the pioneers in the study of instabilities and chaos in lasers, as well as in the theoretical understanding of complex phenomena in nonlinear optical systems.

2.8 Statistical Mechanics, Origin of Irreversibility and Ergodic Theory

In this sense, scientists such as George Birkhoff (1884-1944) stand out, who proposed the ergodic theorem, which was later demonstrated by the German mathematician Eberhard Hopf (1902-1983) using the fact of the ergodicity of trajectories on surfaces of negative curvature constant that the French mathematician Jacques Hadamard (1865-1963) had pointed out a few years before. However, these results had little impact on substantiating nonequilibrium statistical mechanics. Details on the ergodic theory and theorem can be seen in Calvin Moore's insightful article[57], and in its relationship with chaos and randomness in Daniel Ornstein's article[58].

---

[56]H. Haken. Analogy between higher instabilities in fluids and lasers. Physics Letters A 53, 77-78 (1975).

[57]C. Moore. Ergodic theorem, ergodic theory, and statistical mechanics. Proc. Natl. Academic Sci. USA 112, 1907-1911 (2015).

[58]D.S. Ornstein. Ergodic theory, randomness, and "chaos." Science 243 (4888), 182-187 (1989).

The importance of the Lorentz gas, initially proposed by the Dutch physicist Hendrik A. Lorentz (1853-1928) in 1905 as a model for the study of electrical conductivity in metals, lies in its thermodynamic physical properties, in that it is ergodic and has a positive Lyapunov exponent. The great achievement of the Russian-American mathematician Yakov G. Sinai[59], who received the Abel Prize in 2014, was precisely to show the connection between the classic Boltzmann-Gibbs set for an ideal gas and a chaotic Hadamard billiard[60].

Ideas from chaos theory have been used to underpin statistical mechanics, establishing deep connections between the dynamical properties of a system, such as Lyapunov exponents and transport properties. Knowledge of both nonequilibrium statistical mechanics and nonlinear dynamics is essential to understand studies on nonequilibrium states. Despite numerous efforts and apparent new perspectives to support nonequilibrium statistical mechanics based on chaos theory, the extraordinary conceptual difficulties of such an endeavor have so far prevented its achievement.

2.9 Oscillations and Bifurcations. The Path of Nonlinear Oscillators

One of the great paths that have led us to the discovery of chaos is the study of nonlinear oscillators. Among the early pioneers on this path we can find the English physicist John William Strutt, Lord Rayleigh (1842-1919), who was interested in understanding the

---

[59]Y.G. Sinai. Dynamical systems with elastic reflections: Ergodic properties of dispersing billiards. Russian Mathematical Surveys, 25(2), 137–189 (1970).
[60]J. Hadamard. Les surfaces à courbures opposées et leurs lignes géodésiques. Journal de Mathématiques Pures et Appliquées 4, 27-74 (1898).

physics of musical instruments. For these types of systems, an initial approach based on linear oscillators is not effective, since real instruments do not produce a simple tone like a linear oscillator does. Therefore, it is necessary to introduce friction and nonlinear recovery terms. That is, it is necessary to use an elastic force different from that provided by Hooke's law: *ut tensio sic vis*. Rayleigh ingeniously used these elements to create models that explained the sound emitted by musical instruments. In his famous book *The Theory of Sound*[61] published in 1877, Rayleigh introduced a series of fairly general methods such as the notion of limit cycle, which describes a periodic motion of the physical system regardless of its initial conditions. There is also a quadratic nonlinear oscillator model known as the Helmholtz oscillator, named after the German physicist and physiologist Hermann von Helmholtz (1821-1894), who in 1863 published a pioneering monograph on the sensations of tone[62], which constitutes a fundamental work in acoustics and sound perception.

He was followed by others such as the Dutch engineer Balthasar van der Pol (1889-1959), as well as the German engineer Georg Duffing (1861-1944)[63,64] well known for the Duffing oscillator. This model

---

[61] J.W. Strutt Baron Rayleigh. *The Theory of Sound. Vol. I* (London: Macmillan and Co., 1877.)

[62] Hermann von Helmholtz. *Die Lehre von den Tonempfindungen als physiologische Grundlage für die Theorie der Musik [The Study of the Sensations of Tone as a Physiological Foundation for Music Theory]*. (1863).

[63] G. Duffing, *Erzwungene Schwingungen bei veränderlicher Eigenfrequenz und ihre technische Bedeutung* (Vieweg & Sohn, Braunschweig, 1918). (Forced oscillations with changing natural frequencies and their technical importance). K. Worden and H. Worden. Appendix: Translation of Sections from Duffing's Original Book. In The Duffing Equation (eds I. Kovacic and MJ Brennan), 2011.

[64] Ivana Kovacic and Michael J. Brennan, Eds. *The Duffing Equation: Nonlinear Oscillators and their Behavior*. (Wiley, 2011).

is paradigmatic for the study of many phenomena in nonlinear dynamics. The theory was later developed in the late 1940s, just after World War II, by two English mathematicians at the University of Cambridge: Mary L. Cartwright (1900-1998) and John E. Littlewood (1885-1977), who demonstrated that many of the experiments of experimental physicists and many of the conjectures of theoretical physicists were derived directly from the analysis of the differential equations of motion[65]. In fact, these mathematicians had followed the ideas of George Birkhoff, and worked on this problem when they were part of the British program to develop radar during World War II.

The nonlinear school of thought in Russia began with the work of Leonid I. Mandelstam (1879-1944) on nonlinear oscillators. This line of work was continued by Alexander A. Andronov (1901-1952) (Fig. 2.10) and by Lev S. Pontryagin (1908-1988), who introduced the notion of structural stability of a system of equations, a concept associated with that of bifurcations of dynamical systems.

The concept of limit cycle bifurcation that had been suggested by Poincaré in 1892, was demonstrated by Andronov in 1930 and by Hopf in 1940, and is called the Andronov-Hopf bifurcation, although it is usually known as simply the Hopf bifurcation. This school continued later in the 50s and 60s in Gorky, now Nizhni Novgorod, obtaining results parallel to the development of the theory in the West. All nonlinear physics methods were developed under the paradigm of nonlinear oscillators and self-oscillations. Another

---

[65]M.L. Cartwright and J.E. Littlewood, On nonlinear differential equations of the second order, I, Jour. London Math. Soc. 20, 180–189 (1945).

Fig. 2.10 The Russian mathematician Alexander A. Andronov (1901–1952) was one of the pioneers of nonlinear science.

important school on nonlinear thinking in the Soviet Union was the Kiev School of Research on Nonlinear Oscillations which was started by Nikolai M. Krylov (1879-1955) and his student Nikolai N. Bogolyubov (1909-1992) (Fig. 2.11), in the early 1930s. They developed much fundamental work on quasi-periodic solutions for non-autonomous systems and established the discipline of Nonlinear Mechanics as part of Physics. Most of his work was published in the book *Introduction to Nonlinear Mechanics* [66] in Russian. The English version edited and translated by the Russian mathematician Solomon Lefschetz (1884-1973) was published in 1943. Furthermore, in 1946 he began directing the project on nonlinear oscillators at Princeton funded by the newly founded Office of Naval Research.

---

[66]N.M. Krylov and N.N. Bogolyubov. *Introduction to Nonlinear Mechanics* (Ukrainian Academy of Sciences Publishers, Kyiv, 1937). English translation by S. Lefschetz. Introduction to Nonlinear Mechanics. (Princeton University Press, 1943).

Fig. 2.11   Nikolai M. Krylov and Nikolai N. Bogolyubov, who developed the School of Nonlinear Mechanics in Kyiv during the 1930s.

In Japan, the theory of nonlinear oscillators and their applications to radiophysics were developed in the school of the Japanese engineer Chihiro Hayashi (1911-1986) at Kyoto University.   Hayashi made notable contributions to the study of nonlinear oscillators and their practical applications in electrical engineering, publishing his famous book *Nonlinear Oscillations in Physical Systems*[67] in 1964. In 1961 a notable event occurred by the Japanese engineer Yoshisuke Ueda[68], who was a doctoral student of Chihiro Hayashi. Ueda studied the dynamics of several nonlinear oscillators, such as the van der Pol oscillator and the Duffing oscillator, and it is precisely in a particular model of the latter that he apparently first found solutions that we

---

[67] Chihiro Hayashi. *Nonlinear Oscillations in Physical Systems* (Princeton University Press, Princeton, 2014).
[68] Yoshisuke Ueda. *The Road to Chaos II* (Aerial Press, Santa Cruz, CA, 2001).

now designate as chaotic. In a recent publication[69], Yoshisuke Ueda offers a first-person account of the developments and insights that led to his identification of chaotic behavior in dynamical systems.

2.10   The Logistic Map and Population Dynamics

One of the paradigmatic systems where complex dynamics including chaotic behavior is shown is the logistic map. Despite its apparent simplicity, the logistic map derives from the logistic equation, which was introduced in 1838 as a growth model of population dynamics by the Belgian mathematician Pierre François Verhulst (1804-1849) (Fig. 2.12) in his writing *Notice sur la loi que la population poursuit dans son accroissement*[70].

The logistic map was popularized by the Australian Robert M. May (1936-2020) (Fig. 2.13) after the publication of his influential article *Simple Mathematical Models with very Complicated Dynamics*[71] and constitutes one of the paradigms of the chaotic behavior of nonlinear dynamical systems.

Robert May began his scientific career as a physicist[72], but he soon moved to biology becoming one of the pioneers of theoretical ecology, which also led him to become a pioneer of chaos theory.

---

[69]Yoshisuke Ueda. Uncovering the Nature of Chaos. 27 November 2023. https://chaos.amp.i.kyoto-u.ac.jp/en/wp-content/uploads/2023/11/UNCyu.pdf.

[70]P.F. Verhulst. Notice sur la loi que la population suit dans son accroissement. Correspondence Mathematique et Physique (Ghent) 10, 113-121 (1838).

[71]Robert M. May. Simple Mathematical Models with very Complicated Dynamics. Nature 261, 459 (1976).

[72]Miguel A.F. Sanjuán. Lord Robert May of Oxford: an exceptional scientist who sought simplicity from the complex. Spanish Journal of Physics 34(2), 47 (2020). (in Spanish)

Fig. 2.12   Pierre François Verhulst (1804-1849).

The quadratic map, which is very similar to the logistic map, had also been widely studied in other contexts by the Frenchman Gaston Julia (1893-1978), by the Hungarian-American John von Neumann (1903-1957) and by the American Norbert Wiener (1894-1964) and the Polish-American Stanislaw Ulam (1909-1984).

2.11   Period-Doubling Route to Chaos

It is enough to consider one of the simplest chaotic systems, such as the logistic map (Fig. 2.14), to observe one of the most surprising phenomena of chaotic dynamics. The key observation is that, as the value of the governing parameter is gradually varied, the periodicity of the system's solutions undergoes a sequence of bifurcations characterized by a doubling of the period at each stage. That is, we go

Fig. 2.13   Robert M. May (1936–2020), Baron May of Oxford. A physicist and pioneer of theoretical ecology, which led him to contribute to chaos theory.

from having orbits of period 1 to orbits of period 2, 4, 8, 16, 32, 64, 128, 256, 512, 1024, etc.

One of the most influential papers in the field was, without a doubt, the previously cited paper by Edward Lorenz *Deterministic nonperiodic flow*. The American mathematician and physicist at the University of Maryland, Professor James A. Yorke (Fig. 2.15), immediately recognized the implications of such a discovery, as well as its philosophical repercussions, making Lorenz's work known to the scientific community.

He later published the article titled *Period Three Implies Chaos*[73] together with his doctoral student Tien-Yien Li in 1975 where he

---

[73]Tien-Yien Li and James A. Yorke. Period Three Implies Chaos. Am Math Mon 82, 985-992 (1975).

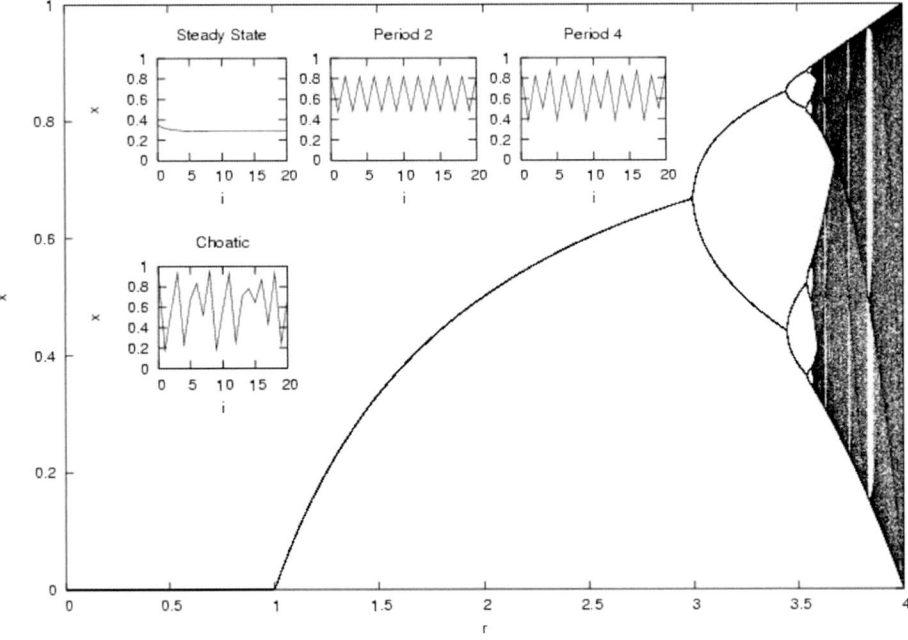

Fig. 2.14   Bifurcation diagram of the logistic map.

introduced the term *chaos* in addition to proving a rigorous theorem. The physicist and mathematician Freeman Dyson wrote an
article about it[74], where he categorically states: "Li-Yorke's article
is one of the immortal gems in mathematical literature [...] The
theorem says that if there is a single orbit with period three, then
chaotic orbits also exist. The test is simple and brief. In my opinion,
this theorem and its proof shed more light than a thousand beautiful pictures on the basic nature of chaos." In 1974, Robert M. May
had made known the phenomenon of period doubling, which he finally published in a famous article published in Nature and which
had an enormous influence on the scientific community. In fact,

---

[74]Freeman Dyson. Birds and frogs. Notices of AMS 56(2), 212-223 (2008).

Fig. 2.15 The American mathematician and physicist James Yorke. He coined the term "chaos" in modern scientific literature.

when Tien-Yien Li and James A. Yorke published their famous article *Period Three implies Chaos*, they credit the period doubling bifurcation cascades to an as yet unpublished work by Robert May. However, in 1975, James Yorke himself became aware of a work by the Ukrainian mathematician Alexander N. Sharkovsky, published in Russian in a Ukrainian mathematics journal in 1964, which contained what is now known as "Sharkovsky's Theorem", which was published in Russian and where part of Li and Yorke's result appeared as a corollary. However, one of the fundamental novelties in Li and Yorke's article is that they showed that the appearance of a period three orbit implied the appearance of all the others, including chaotic orbits, while Sharkovsky did not talk about chaotic orbits.

But the story does not end there. It is here where the figure of the Finnish mathematician Pekka Juhana Myrberg (1892-1976) appears[75], who wrote a series of articles in 1958, 1959, 1962 and 1963, in French and Finnish journals, where he perfectly describes the cascades of bifurcations by period doubling for the quadratic map, with properties similar to the logistic map, using computational numerical simulations. This explains why he found what he called the end of the spectrum, which was later called the Feigenbaum point, that is, the limit where period-doubling bifurcations of powers of two occur.

The concept of renormalization group had been used in the field of statistical mechanics to study so-called critical phenomena and phase transitions and its development in these fields had earned physicist Kenneth Wilson the Physics Nobel Prize in 1982. Subsequently, these methods were applied by the physicist Mitchell Feigenbaum (1944-2019) who discovered the existence of universal critical exponents that characterized the transition from periodic to chaotic motion in one-dimensional maps with the property of period doubling. After the publication of his article in 1978, the Feigenbaum bifurcation diagram, the Feigenbaum point and the Feigenbaum universal constant became popular in the scientific community. Simultaneously, the same discovery was made by the Frenchmen Pierre Coullet and Charles Tresser, who at the time were doctoral students at the University of Nice, and by the German physicists at the University of Marburg, Siegfried Grossmann and Stefan Thomae.

Until the early 1980s, most work was of a theoretical nature or the result of numerical explorations with computers. In any case,

---

[75]https://inaesp.org/Legacy/.

the important consequences that these theoretical discoveries had for physics were always considered, as well as the possible importance for understanding the transition to fluid turbulence. French physicist Albert Libchaber (Fig. 2.16), carried out one of the first experiments demonstrating the period-doubling phenomenon when studying Rayleigh-Bénard convective cells in the late 1970s[76]. The American physicist Robert Shaw of the University of California at Santa Cruz conducted a simple and particularly relevant experiment with a simple dripping faucet[77].

Fig. 2.16   Physicists Albert Libchaber, Harry Swinney, and Jerry Gollub have been pioneers in experimental work on chaos.

Another important experimental milestone was carried out by the American physicists Jerry Gollub and Harry Swinney (Fig. 2.16), who also found the period-doubling phenomenon by reproducing the classic Taylor-Couette experiment of fluid motion. His contributions to the experimental verification of some of the ideas derived from chaos theory have stimulated much experimental work[78].

[76]J. Maurer and A. Libchaber. Rayleigh-Bénard experiment in liquid helium; frequency locking and the onset of turbulence. Journal de Physique Lettres 40 (16), 419-423 (1979).

[77]Robert Shaw. *The dripping faucet as a model chaotic system* (Aerial Press, Santa Cruz, CA, 1984).

[78]J.P. Gollub and H.L. Swinney. Onset of turbulence in a rotating fluid. Physical Review Letters, 35(14), 927–930 (1975). H. L. Swinney and J.P. Gollub. Hydrodynamic instabilities and the transition to turbulence. Physica D: Nonlinear Phenomena 1(3), 260–276 (1978).

2.12   On the Origins of Complexity

At the beginning of the 20th century, fundamental developments occurred in two new fields of research in Physics that represent an enormous conceptual revolution in the development of science. On the one hand, the theory of relativity that helped us understand the world on cosmic scales and quantum mechanics that involved knowledge and exploration of the microscopic world at the atomic and subatomic level. On the other hand, during the second half of the 20th century, we have been able to see how nonlinear dynamics and chaos theory emerged as one of the most fruitful fields of activity in research. Likewise, the discipline of complexity, or the physics of complex systems, has received a great boost, including new lines of research and providing a new way of doing things.

Although for many it remains a relatively new and emerging concept, its widespread recognition and adoption have only gained momentum in recent years, even though its intellectual roots can be traced back to significantly earlier periods. In exploring the foundational ideas and pioneering efforts that have shaped the evolution of this conceptual framework, one that broadly encompasses the multifaceted notion of complexity, it is essential to acknowledge the contributions of American neuroscientist Warren McCulloch[79] (1898–1969). McCulloch, in collaboration with the young and brilliant mathematician Walter Pitts (1923–1969) (Fig. 2.17), introduced in 1943 what is widely regarded as one of the first

---

[79]Tara H. Abraham. *Rebel Genius: Warren S. McCulloch's Transdisciplinary Life in Science* (The MIT Press, Cambridge, MA, 2016).

formal mathematical models of a neuron. This model, known as the McCulloch–Pitts neural model, was a groundbreaking attempt to represent the logical structure of neural activity using mathematical and computational tools, with the goal of shedding light on the basic principles underlying brain function and cognition. This innovative model not only provided a foundational framework for the emerging field of computational neuroscience but also exerted a significant influence on subsequent developments in computer science. Most notably, the eminent polymath John von Neumann later drew upon the McCulloch–Pitts model when formulating the logical design of early digital computers, thereby establishing a conceptual bridge between neurobiology and computer architecture that remains relevant to this day. McCulloch also played a prominent role organizing the Macy Conferences (1946-1953) in the 1940s, which aimed to understand feedback mechanisms in biological, technological and social systems, and constituted one of the first forums to promote interdisciplinarity. Numerous scientists from various disciplines participated in them, among whom we can mention the psychiatrist William Ross Ashby, the anthropologist Gregory Bateson, the mathematicians John von Neumann, Walter Pitts and Norbert Wiener, the biophysicist Max Delbrück, the information theorist Claude Shannon and Warren McCulloch himself as moderator.

Very relevant in relation to the origins of complexity is the recent project of the Santa Fe Institute, which has published 88 articles considered fundamental and commented in four large volumes under the

Fig. 2.17   Walter Pitts (1923-1969) and Warren McCulloch (1898-1969).

title *Foundational Papers in Complexity Science*[80]. The first volume features an excellent introduction to the foundations of complexity science, written by the current director of the Santa Fe Institute, David C. Krakauer. In this essay, he identifies four historical pillars of complexity science: Evolution and Adaptation, Dynamics and Control, Statistical Mechanics and Thermodynamics, and Computation and Logic. Due to its significant interest and contribution to the field, the essay has been published separately as an independent

---

[80] *Foundational Papers in Complexity Science.* Edited by David C. Krakauer. 4 volumes (SFI Press, Santa Fe, NM, 2024).

book, offering readers a compelling and accessible introduction to the foundations of complexity science[81].

2.13 The Advent of Computers and Their Impact on Science

Science has progressed through the application of various methodologies, among which experimental and analytical methods have played undeniably central roles. Nevertheless, the development and understanding of complex systems would be inconceivable without the advent and continued evolution of computational tools. The presence of computers has not only transformed the landscape of scientific inquiry but has also redefined the very way we conceptualize and engage with complexity, ushering in a paradigm shift from which there is no return.

In light of this, it is essential to acknowledge the profound and far-reaching contributions of two towering figures in the history of science, Alan Turing (1912–1954) and John von Neumann (1903–1957), whose foundational work laid the groundwork for modern computing (Figs. 2.18 and 2.19). Their visionary insights and theoretical advances were instrumental in catalyzing the computational revolution that underpins contemporary approaches to complex systems.

On the other hand, it is of special interest to mention the figure of the American scientist Warren Weaver (1894-1978) (Figs. 2.20), who among other things was co-author with Claude Shannon (1916-2001)

---

[81] David C. Krakauer, *The Complex World: An Introduction to the Foundations of Complexity Science* (SFI Press, Santa Fe, NM, 2024).

Fig. 2.18　The IAS Machine was an early 'stored program' computer, named after the Institute for Advanced Studies at Princeton, where it was built over 1946-51 under John von Neumann's supervision.

of the famous book *The Mathematical Theory of Communication*[82] published in 1949.

In 1948, Warren Weaver published a highly influential and forward-looking article titled "Science and Complexity"[83] in American Scientist. Widely regarded as a foundational text, the article was remarkably prescient in its exploration of themes that would later become central to the study of complexity. Many of the concepts that have gained prominence in recent decades were already present in Weaver's insightful analysis, in which he examined the nature and

---

[82] Claude E. Shannon and Warren Weaver. *The Mathematical Theory of Communication* (The University of Illinois Press, 1949).

[83] Warren Weaver. Science and Complexity. American Scientist, 36(4), 536–544 (1948).

Fig. 2.19   Alan Turing (1912-1954) and John von Neumann (1903-1957).

challenges of scientific problems through the lens of complexity, an approach that, even nearly three-quarters of a century later, feels strikingly contemporary.

Weaver distinguishes between simple problems, involving a small number of variables, and organized complexity, characterized by a large number of interacting variables, using the metaphor of predicting the final positions of billiard balls after a break to illustrate the intricate interdependencies involved. He also introduces the idea of disorganized complexity, setting the stage for future developments in systems thinking and complexity science.

Importantly, this pioneering article makes a compelling case for the emerging field of computer science, not only as a technical discipline but as a multidisciplinary framework for problem-solving. Weaver advocates for the use of computational methods

Fig. 2.20   Warren Weaver (1894–1978), a pioneer in the use of computers in scientific research.

and emphasizes the need for interdisciplinary collaboration, anticipating the formation of research groups that integrate diverse fields
to address the multifaceted challenges of complex systems.

Another pivotal figure in the narrative we are constructing
is the mathematician Norbert Wiener (1894–1964) (Fig. 2.21), a
professor at MIT who succeeded in fostering a profoundly interdisciplinary intellectual environment, particularly notable for its numerous applications to the life sciences.

Fig. 2.21   Norbert Wiener (1894–1964), one of the founders of Cybernetics.

Among his many contributions to science, Wiener is best known for founding the field of cybernetics, the core concepts of which were presented in his groundbreaking book *Cybernetics: Or Control and Communication in the Animal and the Machine*[84]. A second edition of the book was published in 1961, only a few years before his death in 1964. It is worth noting that the intellectual roots of cybernetics

---

[84]Norbert Wiener. *Cybernetics: Or Control and Communication in the Animal and the Machine* (Paris: Hermann & Cie, 1948).

can be traced even further back to the Scottish physicist James Clerk Maxwell, a polymath whose diverse interests anticipated some of the foundational ideas of this emerging field.

Wiener's intellectual curiosity spanned a remarkable range of disciplines. His work extended into engineering, particularly in control and communication theory, as well as into computing, statistical mechanics, stochastic processes, cell biology, neuroscience, psychiatry, economics, and the social sciences. In this regard, Wiener was far ahead of his time, engaging deeply with subjects that today are considered integral to the science of complex systems.

He remains an undeniably fascinating and visionary figure, one who, more than seventy years ago, was already grappling with many of the intricate problems that continue to challenge scientists across disciplines today.

Chapter 3

Nobel Prizes in Physics. Physics, Emergence, Nonlinearity, Indeterminism and Life Sciences

I wish to emphasize again that the progress of physics will certainly depend to a large extent on the progress of nonlinear mathematics, of methods of solving nonlinear equations. It may still be that every such problem is individual and requires individual methods. Yet, as I have said, there are definitely some common features and therefore one can learn by comparing different nonlinear problems[85].

— Werner Heisenberg

3.1 Introduction

The remarkable expansion of scientific activity in recent decades has led to an increasing overlap between disciplines, with many fields finding meaningful applications in other areas of research. A prominent example of this interdisciplinary convergence is the application of physics, mathematics, and engineering to the life sciences, encompassing not only biology, but also biomedical sciences and biotechnology. Far from being a recent trend or temporary shift, the influence of these disciplines on the life sciences has deep historical roots. Their contributions have played a significant role in shaping both

---

[85]Werner Heisenberg. Nonlinear problems in physics. Phys. Today 20(5), 27 (1967).

the conceptual foundations and methodological approaches of modern biological research.

Among the mathematical models most widely used in computational neuroscience, particularly those aimed at analyzing the brain as a complex system, the Hodgkin–Huxley model stands out as a foundational contribution. In 1952, Alan L. Hodgkin and Andrew F. Huxley (Fig. 3.1) published a landmark series of five articles[86], in which they described the experiments they conducted to determine the laws governing ionic movement in nerve cells during an action potential.

Fig. 3.1   Alan L. Hodgkin and Andrew F. Huxley were awarded the Nobel Prize in Physiology or Medicine in 1963 for their neuronal model.

[86]A.J. Hopper, H. Beswick-Jones, and A.M. Brown. A color-coded graphical guide to the Hodgkin and Huxley papers. Advances in Physiology Education 46(4), 580-592 (2022).

They developed a mathematical model to explain the behavior of nerve cells in the giant squid. Remarkably, this model was proposed well before the advent of electron microscopes and computer simulations, yet it enabled scientists to gain a basic understanding of how nerve cells function, despite not knowing the detailed behavior of cell membranes at the time. Their research helped solve a long-standing mystery dating back to the famous 1791 experiments by the Italian physicist Luigi Galvani[87], who demonstrated the conduction of electricity in frog muscles. The Hodgkin–Huxley neural model remains a cornerstone in both mathematical and computational neuroscience. In recognition of their groundbreaking work, Alan Hodgkin and Andrew Huxley were awarded the 1963 Nobel Prize in Physiology or Medicine, along with Sir John C. Eccles, for their discoveries concerning the ionic mechanisms involved in excitation and inhibition in the peripheral and central parts of nerve cell membranes.

Another model of particular importance in computational neuroscience is the FitzHugh–Nagumo model (1961), a simplification of the original Hodgkin–Huxley model developed in 1952. While the Hodgkin–Huxley model involves four variables to describe the dynamics of nerve excitation and recovery, the FitzHugh–Nagumo model reduces this to two variables, typically referred to as a *fast* variable and a *slow* variable. This reduction preserves the essential qualitative features of neuronal excitability while significantly simplifying the mathematical complexity. Richard FitzHugh, who was

---

[87]Luigi Galvani. *De viribus electricitatis in motu musculari* (Arnaldo Forni Editore, Bologna, 1998). Originally published in 1791.

working at the National Institutes of Health in Bethesda, Maryland, referred to his formulation as the Bonhoeffer–Van der Pol (BVP) model, as it retained many of the key characteristics of the well-known Van der Pol oscillator. The Van der Pol relaxation oscillator was originally proposed in 1928 by the Dutch engineer Balthasar Van der Pol as a model for electrical activity in the human heart. The model was further advanced by Jinichi Nagumo, from the Department of Applied Physics at the University of Tokyo, who implemented it using an electronic circuit. The FitzHugh–Nagumo model is primarily used to describe the same type of phenomena as the Hodgkin–Huxley model, namely, the control of the electrical action potential across a cell membrane. This process occurs through the modulation of ionic currents across the membrane, which leads to changes in membrane potential that serve as the basis for electrical signal transmission between cells. As noted earlier, the FitzHugh–Nagumo model is not limited to neuronal modeling. It is also widely applied to study excitability in other types of cells, such as cardiac and muscle cells, making it a broadly useful tool in the study of excitable systems across biology.

3.2 Physics, Biology, Complexity

Numerous prominent mathematicians and physicists, including several Nobel Prize winners, have addressed biological problems and explored aspects now central to the science of complexity. Their work demonstrates that, beyond their recognized fields of specialization, many ventured into what we would today classify as interdisciplinary or complexity science.

One of the earliest examples is Norbert Wiener, a pioneer of cybernetics in the 1940s, whose ideas laid the groundwork for systems theory and feedback regulation in both machines and biological organisms. In the 1960s, physicist Jack D. Cowan emerged as a pioneer in the nonlinear mathematical modeling of neural systems. Leon N. Cooper, recipient of the 1973 Nobel Prize in Physics, dedicated much of his later career to the study of the brain. Max Delbrück, who received the 1969 Nobel Prize in Physiology or Medicine, played a pivotal role in the founding of molecular biology.

Other major figures include Nicholas Metropolis, known for coining the term *Monte Carlo methods*, who worked on the MANIAC computer and applied these techniques to problems in computational molecular biology. Leo Szilard, best known for his work in nuclear physics, shifted his focus to biology after World War II and made influential contributions to the field. Additionally, Nobel laureates in Physics such as Erwin Schrödinger, Niels Bohr, Philip W. Anderson, and Murray Gell-Mann also made significant intellectual contributions to the development of complexity science.

In 1943, Erwin Schrödinger, recipient of the 1933 Nobel Prize in Physics and director of the Dublin Institute for Advanced Studies, delivered a series of public lectures titled *What is Life?* These lectures were published the following year as a short book of the same name[88], (Fig. 3.2) which had a profound influence on the emergence of molecular biology and famously inspired the search for the structure of DNA.

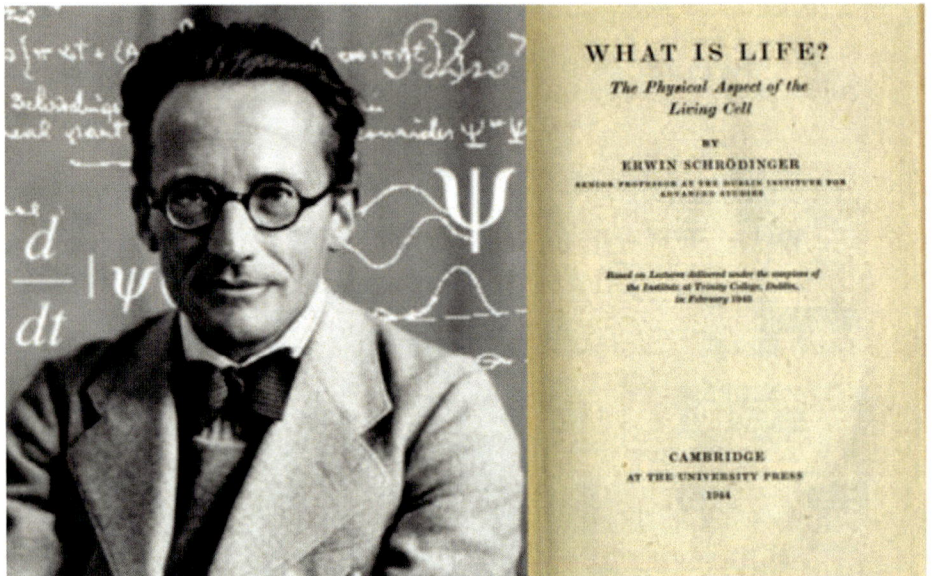

Fig. 3.2   Erwin Schrödinger, Nobel Prize in Physics in 1933, best known for his contributions to quantum mechanics. Author of the influential book "What is Life?" published in 1944.

3.3   Physics and Emergence

One of the fundamental concepts in the study of complexity is the notion of emergence. In the field of physics, one can observe numerous examples of systems where emergent properties are evident, as in the case of superconductivity and superfluidity. Importantly, there is fundamental research dedicated to exploring complex phenomena, where the predominant approach is not reductionism, which has been the main paradigm in the evolution of physics in recent years, but

---

[88]Erwin Schrödinger, *What is Life?   The Physical Aspect of the Living Cell*, Cambridge University Press, 1944.

rather emergence. An essential aspect is that these emergent complex phenomena do not follow directly from the underlying microscopic laws, although, of course, they comply with them.

Some of these ideas were masterfully presented by the American physicist Philip W. Anderson (1923-2020)[89] (Fig. 3.3), Nobel Prize in Physics in 1977, in an article published in Science magazine in 1972 and titled *More is different*[90], where he makes the idea very clear that: "At each level of complexity entirely new properties appear, and the understanding of the new behaviors requires research which I think is as fundamental in its nature as any other."

Fig. 3.3   Philip W. Anderson (1923–2020), Nobel Prize in Physics 1977.

---

[89] Andrew Zangwill. *A Mind Over Matter: Philip Anderson and the Physics of the Very Many* (Oxford University Press, 2021).

[90] P.W. Anderson. More is different. Science 177, 393-396 (1972).

Philip W. Anderson introduces some aspects of the physics of complex systems in the article entitled *Physics: The Opening to Complexity*[91], where referring to the work of physicists in complexity he points out: "But another large fraction are engaged in an entirely different type of fundamental research: research into phenomena that are too complex to be analyzed straightforwardly by simple application of the fundamental laws. These physicists are working at another frontier between the mysterious and the understood: the frontier of complexity. At this frontier, the watchword is not reductionism but emergence. Emergent complex phenomena are by no means in violation of the microscopic laws, but they do not appear as logically consequent on these laws."

He also expresses very well the use of models in physics, where they must necessarily be simplified and contain the essence of the problem under study: "Very often such a simplified model throws more light on the real workings of nature than any number of *ab initio* calculations of individual situations, which even where correct often contain so much detail as to conceal rather than reveal reality. It can be a disadvantage rather than an advantage to be able to compute or to measure too accurately, since often what one measures or computes is irrelevant in terms of mechanism. After all, the perfect computation simply reproduces Nature, it does not explain her."[92]

---

[91]P.W. Anderson. Physics: The Opening to Complexity. Proceedings of the National Academy of Sciences of the USA 92, 6653-6654 (1995).
[92]P.W. Anderson. Local Moments and Localized States. Reviews of Modern Physics 50, 191–201 (1978).

It is also of great interest to read what he says about the role of complexity in physics in an essay published in Physics Today: "If broken symmetry, localization, fractals and strange attractors are not "fundamental," what are they? A movement is under way toward joining together into a general subject all the various ideas about ways new properties emerge. We call this subject the science of complexity. Within this topic, ideas equal in depth and interest to those in physics come from some of the other sciences. This movement is overdue and healthy. On the other hand, one may well be apprehensive—or at least I am—that such an enterprise might go the way of General Semantics, General Systems Theory and other well-meant but premature and intellectually lightweight attempts at building an overall structure. We complexity enthusiasts (perish the thought that we be called complexity scientists!) are talking, at least for the most part, about specific, testable schemes and specific mechanisms and concepts. Occasionally we find that these schemes and concepts bridge subjects, but if we value our integrity, we do not attempt to force the integration. A number of institutions have grown up to foster this kind of work, and in a future column I'd like to write about one of them, the Santa Fe Institute. Meanwhile, let me give my own answers to my opening questions: Complexity, as defined in the preceding paragraph, is often physics. It is the leading edge of science. And it is surely exhilarating."[93] Precisely one of the key ideas in the methodology for studying complex systems is to

---

[93]P.W. Anderson. Is Complexity Physics? Is It Science? What is It? Physics Today 44(7), 9–11 (1991).

look for the essential issues that allow us to understand a problem without needing to understand all the minor details it contains. A pictorial representation of this idea is reflected by Pablo Picasso's brilliant lithograph of the bull (Fig. 3.4). It is the same idea that is hidden when we look at a model in physics and compare it with the real object.

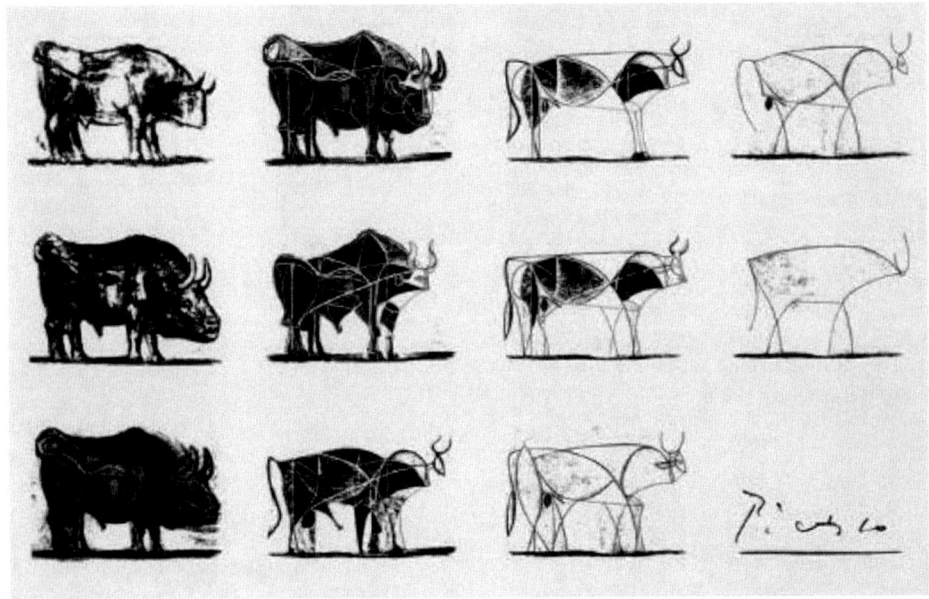

Fig. 3.4   Pablo Picasso distills the essence of a bull in his famous lithograph "The Bull" (1945).

3.4   The End of Reductionism: The Emergence

In relation to the physics of emergence, it is also worth mentioning Robert Laughlin, Nobel Prize winner in Physics in 1998, who used to propose to his best students at Stanford University the problem of deducing the laws of superfluidity from first principles, knowing that it is impossible. Precisely to show you the importance of emergent properties in physics, which is the fundamental argument of his book *A Different Universe: Reinventing physics from the bottom down*[94] with two magnificent reviews by Nobel Prize winner in Physics Anthony Leggett[95] and Phil Anderson[96].

The book is based on an interesting article titled *The Theory of Everything*[97], where among the many issues he points out we can highlight the following two paragraphs: "The central task of theoretical physics in our time is no longer to write down the ultimate equations but rather to catalogue and understand emergent behavior in its many guises, including potentially life itself. We call this physics of the next century the study of complex adaptive matter. For better or worse we are now witnessing a transition from the science of the past, so intimately linked to reductionism, to the study of complex adaptive matter, firmly based in experiment, with its hope for providing a jumping-off point for new discoveries, new concepts, and new wisdom." [...] "The end of Reductionism, for it is actually a call to those of us concerned with the health of physical science to face the truth that in most respects the reductionist ideal has reached its limits as a guiding principle. Rather than a Theory of Everything we appear to face a hierarchy of Theories of Things, each emerging from its parent and evolving into its children as the energy

---

[94]R. Laughlin. *A Different Universe: Reinventing Physics from the bottom down* (Basic Books, 2005).
[95]A. Leggett. A Different Universe: Reinventing Physics from the Bottom Down. Physics Today 58 (10), 77–78 (2005).
[96]P.W. Anderson. Emerging physics: a fresh approach to viewing the complexity of the universe. Nature 434, 701, (2005).
[97]R.B. Laughlin and D. Pines. The Theory of Everything. Proc. Natl. Academic Sci. USA 97, 28-31 (2000).

scale is lowered. The end of reductionism is, however, not the end of science, or even the end of theoretical physics."

3.5   Physics, Nonlinearity and Indeterminism

Due to the enormous consequences on determinism in physics that quantum mechanics has brought with it through the Heisenberg uncertainty principle, the idea of indeterminism has been directly related to quantum mechanics. This has somehow led to considering classical mechanics as completely deterministic and predictable, which is not entirely true[98].

3.5.1   *Max Born and classical determinism*

Once again, the same ideas that we had discussed about James Clerk Maxwell emerge, that is, chaos and determinism and the idea of sensitive dependence on initial conditions, which is the hallmark of chaos. It is fascinating to corroborate that the idea of sensitive dependence on initial conditions was considered in detail by the German physicist Max Born (1882-1970), Nobel Prize winner in Physics in 1954, in an article entitled *Is classical mechanics in fact deterministic?*[99]. He presented in this paper a study of a two-dimensional Lorentz gas initially proposed by the Dutch physicist Hendrik A. Lorentz (1853-1928) in 1905 as a model for the study of electrical conductivity in metals. In this model, a particle moves in

---

[98]An interesting and recent discussion of classical indeterminism in the context of quantum mechanics can be seen in: Manabendra Nath Bera et al. Randomness in quantum mechanics: philosophy, physics and technology. Rep. Prog. Phys. 80, 124001 (2017).
[99]Max Born. Is Classical Mechanics in Fact Deterministic? Phys. Blätter 11(9), 49 (1955).

a plane filled with hard spheres and collides with them, so a small change in the initial conditions will significantly alter the particle's trajectory. This fact led Born to conclude that the determinism traditionally related to classical mechanics is not real, since it is not possible to know with infinite precision the initial conditions of a physical experiment.

Furthermore, in the speech[100], which he delivered on the occasion of the awarding of the Nobel Prize in 1954, the following words appear: "Newtonian mechanics is deterministic in the following sense: If the initial state (positions and velocities of all particles) of a system is accurately given, then the state at any other time (earlier or later) can be calculated from the laws of mechanics. All the other branches of classical physics have been built up according to this model. Mechanical determinism gradually became a kind of article of faith: the world as a machine, an automaton. As far as I can see, this idea has no forerunners in ancient and medieval philosophy. The idea is a product of the immense success of Newtonian mechanics, particularly in astronomy. In the 19th century it became a basic philosophical principle for the whole of exact science. I asked myself whether this was really justified. Can absolute predictions really be made for all time on the basis of the classical equations of motion? It can easily be seen, by simple examples, that this is only the case when the possibility of absolutely exact measurement (of position, velocity, or other quantities) is assumed. Let us think of a particle

---

[100]Max Born. Nobel Lecture. The Statistical Interpretations of Quantum Mechanics. Nobel Lecture, December 11, 1954.

moving without friction on a straight line between two end-points (walls), at which it experiences completely elastic recoil. It moves with constant speed equal to its initial speed v_0 backwards and forwards, and it can be stated exactly where it will be at a given time provided that v_0 is accurately known. But if a small inaccuracy Δv_0 is allowed, then the inaccuracy of prediction of the position at time t is $t\Delta v_0$ which increases with t. If one waits long enough until time $t_c = l/\Delta v_0$ where l is the distance between the elastic walls, the inaccuracy Δx will have become equal to the whole space l. Thus it is impossible to forecast anything about the position at a time which is later than t_c. Thus determinism lapses completely into indeterminism as soon as the slightest inaccuracy in the data on velocity is permitted."

3.5.2   *Feynman and chaos theory*

The American physicist Richard Feynman (1918-1988), winner of the Nobel Prize in Physics in 1965 (Fig. 3.5), makes similar reflections in his well-known book *The Feynman Lectures on Physics*[101], where he explains that indeterminism does not belong exclusively to quantum mechanics; it is a basic property of many physical systems.

In section 38-6 of the first volume, entitled *Philosophical Implications*, a masterful description of indeterminism in classical mechanics is made, and in particular of one of the most important aspects

---

[101]R.P. Feynman, R.B. Leighton, and M. Sands. *The Feynman Lectures on Physics*. Vol. *I Mainly Mechanics, Radiation and Heat* (Addison-Wesley, Reading, Massachusetts, 1963). Section 38-6 entitled "Philosophical implications", corresponding to chapter 38.

Fig. 3.5   Richard Feynman (1918–1988), Nobel Prize in Physics in 1965.

in chaos theory: sensitive dependence on initial conditions, which results in chaotic systems being unpredictable in the long term. Finally stating: "For already in classical mechanics there was indeterminability from a practical point of view."

3.5.3   *Enrico Fermi (1901-1954) and the FPUT experiment*

Enrico Fermi (1901-1954), Nobel Prize winner in Physics in 1938, is well known for his work on radioactivity, as well as nuclear physics. Less well known is the role he also played in the dawn of nonlinear physics. In fact, for work carried out a year before his death, he is recognized as one of the pioneers of numerical computer simulation for a famous experiment he performed at Los Alamos in 1954 known as the Fermi-Pasta-Ulam-Tsingou experiment[102] in which Enrico Fermi, John Robert Pasta, Stanisław Ulam and Mary Tsingou participated. (Fig. 3.6)

Fig. 3.6   Enrico Fermi, John Pasta, Stanisław Ulam and Mary Tsingou.

Mary Tsingou, along with a group of recent graduates, was hired to carry out manual calculations. After being assigned to the theoretical division, she quickly moved to the division led by Nicholas Metropolis, to work on the new MANIAC I that no one knew how to program. She and Mary Hunt were the first programmers to do exploratory work. The computer was mainly used for

---

[102]E. Fermi, J. Pasta, S. Ulam. Studies of Nonlinear Problems. Los Alamos National Laboratory Document LA-1940 (1955). in E. Segrè, ed., Collected Papers of Enrico Fermi, Vol. 2, (U. Chicago Press, Chicago, 1965).

weapons-related tasks, but from time to time, and especially on weekends, researchers could use it to study physics problems. Mary Tsingou and John Pasta were the first to create computer graphics when they considered an explosion problem and visualized it on an oscilloscope.

Fermi cleverly proposed that computers could be used to perform numerical experiments, so he proposed to verify statistical physics' prediction of the thermalization of solids. Preliminary calculations confirmed that energy introduced in a single Fourier mode shifts to other modes, so the initial quasi-periodic behavior was not observed due to the slowness of the computer. However, one day, the computer did not stop as planned, and the researchers discovered to their astonishment that almost all the power returned to the initial mode, recovering the original state almost perfectly. Young Mary Tsingou[103] performed the calculations for this first numerical experiment at MANIAC I in 1955 to simulate energy relaxation.

An excellent overview of the significance of the FPUT experiment in the development of Nonlinear Dynamics and Statistical Physics is presented in the monograph *The Genesis of Simulation in Dynamics* by Thomas P. Weissert[104]. This work also offers a compelling discussion connecting the FPUT experiment with the KAM theorem, the Hénon-Heiles Hamiltonian, and Melnikov theory, all of which were mentioned earlier.

---

[103]Thierry Dauxois. Fermi, Pasta, Ulam, and a mysterious lady. Physics Today 61(1), 55–57 (2008).
[104]Thomas P. Weissert, *The Genesis of Simulation in Dynamics: Pursuing the Fermi-Pasta-Ulam Problem* (Springer, NY, 1997).

It is not surprising that prominent physicists, such as the Italian Angelo Vulpiani, affirm that the majority of physicists recognize the importance of nonlinear systems not so much because of the theorems of Kolmogorov, Arnold and Moser, but because of the suggestive power of simulation graphs of the pioneers of chaos like Fermi, Pasta, Ulam, Chirikov, Lorenz or Hénon[105].

3.5.4   *Erwin Schrödinger (1887-1961)*

Erwin Schrödinger's scientific life was very multifaceted, and among his writings there are direct or indirect contributions to ideas about nonlinearity and indeterminism, in addition to those already mentioned in the life sciences.   In 1935 he published a brief work in Spanish in the journal Annals of the Society of Physics and Chemistry, entitled: *Are the true equations of the electromagnetic field linear?*[106], where he considers that it is still curious that while the pressure of light is obtained from Maxwell's equations, which are linear; sound pressure is obtained by a linear approximation of the true equations of hydrodynamics, which are nonlinear.   In this way Schrödinger thinks that Maxwell's equations are nothing more than the linear approximation of the true equations of electromagnetism, which by analogy must have been nonlinear.

The following quotes show the extent to which ideas so closely related to nonlinearity and indeterminism and so natural in the later developments of nonlinear dynamics and chaos theory were in

---

[105]Massimo Falcioni and Angelo Vulpiani. Enrico Fermis's contribution to non-linear systems: The influence of an unpublished article, pp 271-285, in Enrico Fermi His Work and Legacy, edited by Carlo Bernardini and Luisa Bonolis (Springer, Berlin, 2001).

[106]Erwin Rudolf Alexander Schrödinger. Are the true equations of the electromagnetic field linear? Annals of the Society of Physics and Chemistry, Vol. 33, 511-517 (1935). (in Spanish)

Schrödinger's mind. As Walter Moore quotes[107], Schrödinger wrote a letter to Max Born saying: "If everything were linear, nothing would influence nothing, said Einstein once to me. That is actually so. The champions of linearity must allow zero-order terms, like the right side of the Poisson equation, $\Delta V = -4\pi\rho$. Einstein likes to call these zero-order terms 'asylum ignorantiae'."

And regarding indeterminism, long before it was formulated in quantum mechanics, he said: "The modem attempts to relinquish determinism are rendered particularly interesting by the fact that their claims with regard to the absence of determinism, far from being vague and inaccurate, are quantitatively quite definite and can be expressed in centimeters, grams and seconds. As a simple example, we may take a mass point in motion either in a state of isolation from others or as a member of a system of many mass points exerting force upon each other. The claim which is made is that its movement cannot be foretold with complete accuracy because, among other things, it would be necessary to know its position and velocity at the initial point of time; and it is claimed that it is impossible in principle to determine both of these exactly." [...]. "Long before modern quantum mechanics made its quantitative statements with respect to the degree of inaccuracy, it was possible, although it was not necessary, to doubt the justification of determinism from a far more general point of view. In fact, such doubts were raised in 1918 by Franz Exner, nine years before Heisenberg set up his relation of indeterminacy. Little attention was paid to them, however, and if

---

[107]Walter Moore. *Schrödinger. Life and thought* (Cambridge University Press, Cambridge, 1989), p. 381. E. Schrödinger. Letter to Max Born, 1942.7.27. Staatsbibliothek Preussischer Kulturbesitz, Berlin, Born Nachlass 704.

support was given to them, as by the author in his inaugural dissertation at Zurich, they met with considerable shaking of heads."[108]

3.5.5 *Werner Heisenberg (1901-1976)*

Werner Heisenberg (Fig. 3.7) is recognized for his important contributions to quantum mechanics. However, his ideas on nonlinear problems in physics, as reflected in his article on this topic[109], are quite unknown. In this article published in Physics Today in 1967 he glossed in detail the relevance of nonlinear problems in physics.

Fig. 3.7   Werner Heisenberg (1901-1976).

---

[108]Indeterminism in Physics. Schrödinger presented this paper on 16th June, 1931 to the Congress of A Society for Philosophical Instruction in Berlin. (Erwin Schrödinger, 1935, Science and the Human Temperament, Chapter III.) pp. 43-65.

[109]Werner Heisenberg. Nonlinear problems in physics. Physics Today 20(5), 27 (1967).

In fact, Isaac Newton already formulated the nonlinear problems of physics because in fact Newton's equations involve nonlinear mathematical equations. His emphasis on nonlinearity led him to write that "the largest part of theoretical physics is devoted to nonlinear problems." In addition, he refers to everything from work on turbulence in fluids to nonlinear formulations of quantum mechanics. His comments on the unpredictability of physical systems affect topics such as the three-body problem of such relevance in celestial mechanics, planetary systems and astrophysics in general, as well as with chaos theory.

Another important aspect very familiar with modern nonlinear dynamical systems theory is when it states that "nonlinear problems have a certain kind of unpredictability. One doesn't know how the solutions will behave after a very long time; I think that this may be a very general feature of nonlinear problems."

Many of the ideas he exposes are particularly clear-sighted, such as: "Finally I wish to emphasize again that the progress of physics certainly will depend to a large extent on the progress of nonlinear mathematics, of methods of solving nonlinear equations. It may still be that every such problem is individual and requires individual methods. Yet, as I have said, there are definitely some common features and therefore one can learn by comparing different nonlinear problems."

3.5.6 *Murray Gell-Mann (1929-2019)*

Murray Gell-Mann[110] was a brilliant theoretical physicist best known for introducing the concept of quarks, the fundamental constituents of protons and neutrons, a groundbreaking insight that earned him the Nobel Prize in Physics in 1969. Beyond his pivotal contributions to particle physics, Gell-Mann also played a key role in advancing the study of complex systems. He was one of the founders of the Santa Fe Institute, a multidisciplinary research center devoted to understanding complexity across disciplines, from physics and biology to economics and social systems.

A major influence, and perhaps the reason many physicists became aware of his engagement with complex systems, was his book *The Quark and the Jaguar*[111]. In *The essence of emergence*[112], British physicist Michael Berry offers a concise overview of the book, highlighting several key themes. These include the notion of complex adaptive systems, the role of chaos and algorithmic complexity, and how simple underlying laws can give rise to emergent structures.

Gell-Mann's work helped bridge the gap between physics and other sciences, emphasizing that complexity is a universal feature of many natural and artificial systems. His interdisciplinary vision has had a lasting impact on the way scientists understand the emergence of order from chaos.

---

[110]George Johnson. *Strange Beauty: Murray Gell-Mann and the Revolution in Physics* (SFI Press, 2nd edition, 2023).

[111]Murray Gell-Mann. *The Quark and the Jaguar: Adventures in the Simple and the Complex* (W. H. Freeman/Little, 1994.

[112]Michael Berry. The essence of emergence. Nature 369, 529 (1994).

Fig. 3.8   Murray Gell-Mann (1929-2019). Physics Nobel Prize 1969.

Finally, I want to highlight the Italian theoretical physicist Giorgio Parisi (1948-) (Fig. 3.9), worthy of the Nobel Prize in Physics in 2021 for his revolutionary contributions to the theory of disordered and random phenomena.

His work has been fundamental in solving numerous problems in the theory of complex systems. Among his achievements, research on spin glasses has been especially notable, opening new perspectives in the understanding of these disordered materials and their

Fig. 3.9   Giorgio Parisi. Physics Nobel Prize 2021.

magnetic properties. This is a category of disordered materials, where Parisi made the crucial breakthrough that allowed the models to be resolved and, more importantly, understood. The theory of spin glasses, and more generally of disordered systems, is an active field of research whose basic model is based on the Ising model that Sam Edwards and Phil Anderson had developed in the 1970s. Parisi is also recognized for his concept of breaking the symmetry of replicas, which has had a profound impact on the field. In addition, he has made important contributions to the study of randomness, including the application of stochastic resonance to climate and the analysis of the multifractal nature of turbulence, as well as understanding the effects of fluctuations.

Related to the work of Parisi is the work of John Hopfield and and Geoffrey Hinton (Fig. 3.10), 2024 Physics Nobel Prize winners "for foundational discoveries and inventions that enable machine learning with artificial neural networks" as the Royal Swedish Academy of Sciences declared. In the document of scientific background to the Nobel Prize in Physics 2024 the Nobel Committee for Physics mentions the work carried out as early as 1943, by Warren McCulloch and Walter Pitts that I have discussed in the beginning of this chapter, where they proposed a model for how the neurons in the brain cooperate[113].

Fig. 3.10   John Hopfield and Geoffrey Hinton. 2024 Physics Nobel Prize winners.

3.6   The University of Maryland Chaos Group

Many basic developments in chaos theory have been developed within the chaos group at the University of Maryland led by James A. Yorke (Fig. 3.11) together with Edward Ott and Celso Grebogi

---

[113]The scientific background to the Nobel Prize in Physics 2024 "For foundational discoveries and inventions that enable machine learning with artificial neural networks". The Nobel Committee for Physics.   https://www.nobelprize.org/uploads/2024/11/advanced-physicsprize2024-3.pdf.

and numerous collaborators who have had an enormous influence on the development of the field with applications in various areas of science and engineering.

Fig. 3.11   James A. Yorke at the Doctor Honoris Causa Celebration at the URJC in 2014.

Some of the groundbreaking ideas and fundamental developments are: chaos control, fractal and Wada structures, transient chaos, nonchaotic strange attractors, as well as rigorous developments of numerical simulations of chaotic orbits. In 1990, the seminal article *Controlling Chaos*[114] appeared, which is known as the OGY method of chaos control, where the authors showed how an adequate strategic

---

[114]E. Ott, C. Grebogi, and J.A. Yorke. Controlling Chaos, Phys. Rev. Lett. 64, 1196 (1990).

manipulation can stabilize chaotic systems by converting apparent randomness into predictable behaviors. It is one of the most cited articles in the chaos literature, if not the most cited. Also using chaos to direct trajectories to precise targets of a dynamical system[115]. A review article of these influential ideas was published in Nature[116].

A basin of attraction is defined as the set of initial conditions whose trajectories go towards a specific attractor. In multistable systems, trajectories can have different destinations due to a small perturbation or uncertainty in the initial conditions. If there are several attractors in a region of phase space, then there will be several basins that will be separated by their corresponding boundaries, which in turn may be formed by smooth curves or fractals[117]. Things could get more complicated in the case of the Wada basins[118], or in the so-called riddled basins[119]. The most important results appeared in separate articles[120] in Science in 1987 and in 1996.

The idea of crisis[121] has to do with the notion of the appearance of sudden qualitative changes of chaotic dynamics. And the key

[115]T. Shinbrot, E. Ott, C. Grebogi, J.A. Yorke. Using chaos to direct trajectories to targets. Phys. Rev. Lett. 65(26), 3215 (1990).

[116]T. Shinbrot, C. Grebogi, JA Yorke, E. Ott. Using small perturbations to control chaos. Nature 363 (6428), 411-417 (1993).

[117]S.W. McDonald, C. Grebogi, E. Ott, J.A. Yorke. Fractal basin boundaries. Physica D: Nonlinear Phenomena 17 (2), 125-153 (1985).

[118]J. Kennedy, J.A. Yorke. Basins of Wada. Physica D: Nonlinear Phenomena 51(1-3), 213-225 (1991). H.E. Nusse, J.A. Yorke. Wada basin boundaries and basin cells. Physica D: Nonlinear Phenomena 90(3), 242-261 (1996).

[119]J.C. Alexander, J.A. Yorke, Z. You, I. Kan. Riddled basins. Int J Bifurcat Chaos 2, 795-813 (1992).

[120]C. Grebogi, E. Ott, J.A. Yorke. Chaos, strange attractors, and fractal basin boundaries in nonlinear dynamics. Science 238 (4827), 632-638 (1987). H.E. Nusse and J.A. Yorke. Basins of attraction. Science 271, 1376-1380 (1996).

[121]C. Grebogi, E. Ott, J.A. Yorke. Chaotic attractors in crisis. Phys. Rev. Lett. 48(22), 1507 (1982).

is to analyze the causes and properties that generate this type of growth or sudden disappearance of the attractors. Associated with a type of crisis called a boundary crisis is the notion of transient chaos[122], which unlike permanent chaos only lasts a finite time. This important concept has given rise to numerous lines of research within a broader context of what is known as transient dynamics, whose field of action is also applicable to numerous scientific disciplines.

Another idea developed by the Maryland group is the idea of nonchaotic strange attractors[123], which has recently had repercussions in Astrophysics with the discovery of the strange nonchaotic stars that I will refer to later. These are a type of attractors that have a fractal dimension, but are not chaotic. Finally, another topic of special importance and very relevant in the development of computational methods has been the study of rigorous results of numerical simulations of chaotic orbits[124].

For more than thirty years, I have maintained a fruitful collaboration with Prof. James Yorke, sharing ideas through numerous joint projects and meetings in various locations. In addition, Prof. James Yorke has been awarded an Honorary Doctorate at Universidad Rey Juan Carlos (URJC), Madrid, Spain, where I had the honor of acting as his sponsor. Furthermore, he has been elected

---

[122]C. Grebogi, E. Ott, J.A. Yorke. Crises, sudden changes in chaotic attractors, and transient chaos. Physica D: Nonlinear Phenomena 7 (1-3), 181-200 (1983).

[123]C. Grebogi, E. Ott, S. Pelikan, J.A. Yorke. Strange attractors that are not chaotic. Physica D: Nonlinear Phenomena 13, 261-268 (1984).

[124]S.M. Hammel, J.A. Yorke and C. Grebogi, Numerical orbits of chaotic processes represent true orbits. Bull. Amer. Math. Soc. 19, 465-469 (1988). T. Sauer, J.A. Yorke. Rigorous verification of trajectories for the computer simulation of dynamical systems. Nonlinearity 4, 961 (1991). T. Sauer, C. Grebogi, J.A. Yorke. How long do numerical chaotic solutions remain valid? Phys. Rev. Lett. 79, 59 (1997).

a Foreign Member at the Royal Academy of Sciences of Spain. In 2003, he received the prestigious Japan Prize in Complexity Science and Technology awarded by the Government of Japan, presented personally by the Emperor of Japan, at whose ceremony I had the honor of being present.

3.7 Interdisciplinary Research in Nonlinear Dynamics, Chaos and Complex Systems

I would like to highlight some of the main contributions to the field made by the Nonlinear Dynamics, Chaos Theory, and Complex Systems Group at the Universidad Rey Juan Carlos, which I lead. My scientific work in this area has been guided by two fundamental principles: interdisciplinarity and internationalization. Nonlinear dynamics is inherently interdisciplinary and multidisciplinary from a methodological standpoint. This means that similar methods and tools can be applied to a wide range of research problems across various disciplines. At the same time, my research has a strong international dimension, as reflected by my collaborations with more than two hundred collaborators of more than twenty-five different nationalities. Below, I will briefly describe some of these contributions, selected based on their impact and thematic diversity, organized through a simple classification[125].

Dynamical systems and fractal structures. This interdisciplinary research covers fields ranging from developments in applied

---

[125]An appendix at the end of the book lists references to the major publications I refer to in this brief review of my contributions to the field.

mathematics and dynamical systems, studies on fractal structures in the geometrical phase space, development of new methods for predicting dynamical systems such as basin entropy, new advances in the study of the Wada basins (Fig. 3.12), as well as new methods for their detection, and bifurcation techniques with delays.

Celestial mechanics and astrophysics. These are various applications in astronomy and astrophysics, and in particular problems of celestial mechanics through the use of paradigmatic Hamiltonian systems such as the Hénon-Heiles Hamiltonian, dynamics of fractal structures of binary black hole shadows, as well as predictability of chaotic dynamics and its application in galactic models.

Applications in physics. Nonlinear dynamics methods have been applied to various problems in physics, such as chaos and quantum entanglement, the study of chaotic dynamics and fractal structures in experiments with cold atoms, new advances in chaotic scattering phenomena, both classical and relativistic, as well as various applications to plasma physics problems.

Control of chaotic systems. After the pioneering work of the Maryland group and especially the widely known OGY chaos control method, other methods have been developed. One of the main contributions in this field is the development of the partial control method, which is basically applied to problems with transient chaos in the presence of fluctuations where the objective of control is to avoid an unwanted but inevitable situation in the dynamics of the system. Likewise, we have advanced in the development of applications of the phase control method, which consists of using the phase of an external periodic disturbance to control chaotic dynamics.

Noise and stochastic phenomena. These investigations contemplate the effects of fluctuations, which are modeled through noise and other stochastic phenomena in dynamical systems of a physical, mechanical or biological nature.

Machine learning, AI and Evolutionary Game Theory. Applications to chaotic dynamics problems where the effectiveness of machine learning algorithms for the classification of basins of attraction has been demonstrated, as well as explorations of chaotic systems in Evolutionary Game Theory.

Nonlinear mechanics. In this context, numerous studies of the dynamics of nonlinear oscillators have been addressed, fundamental in the modeling of numerous problems in science, as well as the development of the phenomenon of vibrational resonance and the study of other nonlinear resonances.

Biological phenomena. Modeling biological systems: developments in computational neuroscience, especially through the use of discrete dynamical system models; Physics and modeling of cancer dynamics through dynamic systems: tumor control and growth. Likewise, we have studied models of ecological complexity, modeling of genetic networks, as well as control of infectious diseases.

Fundamentals of nonlinear dynamics and chaos theory. These are contributions of a more mathematical nature addressing the analysis of issues of a topological nature associated with the so-called indecomposable continua, related to the topological structure of chaos; as well as the concept of hetero-chaos as a new paradigm in the study of high-dimensional dynamic systems.

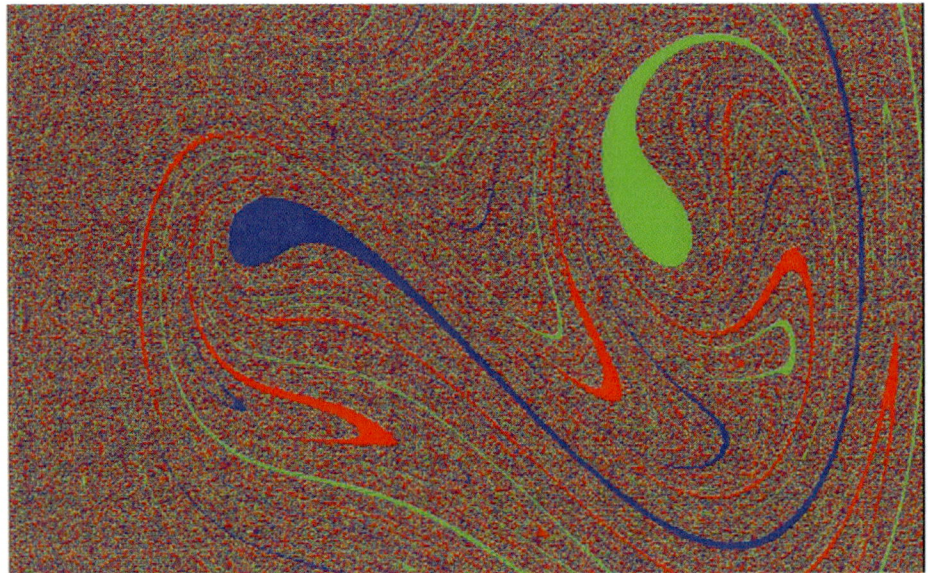

Fig. 3.12   Wada basins of attraction of the nonlinear Duffing oscillator. Each color indicates the attractor to which each initial condition will converge. The figure shows the high final uncertainty of each point in this region of phase space. J. Aguirre and M.A.F. Sanjuán. Unpredictable behavior in the Duffing oscillator: Wada basins. Physica D: Nonlinear Phenomena 17, 41-51 (2002).

Chapter 4

Interdisciplinarity in Sciences

This frontier of complexity is by far the most active growth point of physics. Physicists are also finding themselves, more and more, working side by side with other scientists in interdisciplinary collaborations at this frontier[126].

— Philipp W. Anderson

4.1 Introduction

As has become clear throughout the text, the influence of the ideas of nonlinearity, chaos and complexity have been invading all scientific disciplines. So, now I would like to make it clear through some relevant examples in the different disciplines (Fig. 4.1) that make up the three sections of the Royal Academy of Sciences of Spain, that is, Mathematics, Physics and Chemistry, Natural Sciences and related subjects.

---

[126]P.W. Anderson. Physics: The Opening to Complexity. Proceedings of the National Academy of Sciences of the USA 92, 6653-6654 (1995).

Fig. 4.1   Interdisciplinarity in the sciences.

4.2   Applied Mathematics, Computer Science and Algorithms

The applications of nonlinear science within mathematics are vast and, unsurprisingly, profoundly influential. Among the many noteworthy contributions, several described by the French mathematician René Lozi deserve special mention[127]. In his exploration of chaotic attractors, Lozi examines whether these phenomena are merely mathematical curiosities or whether they contribute meaningfully to scientific progress.

He outlines a broad array of practical applications, ranging from cryptography and metaheuristic optimization to algorithms such as

---

[127]R. Lozi. Are Chaotic Attractors Just a Mathematical Curiosity or Do They Contribute to the Advancement of Science?, Chaos Theory and Applications, 5(3), 133–140 (2023).

Particle Swarm Optimization (PSO), Differential Evolution (DE), and the Self-Organized Migrating Algorithm (SOMA). Additionally, Lozi highlights their significance in areas like chaotic communications and the development of memristors, an innovation attributed to the renowned Chinese-American engineer Leon Chua.

4.3 Chaos and Complexity in Physics

4.3.1 *Astronomy and Astrophysics*

One of the pioneering figures in the application of nonlinear dynamics and chaos theory to astronomy is the Greek astronomer Georges Contopoulos. His foundational contributions are thoroughly presented in his seminal work *Order and Chaos in Dynamical Astronomy*[128], as well as in his scientific autobiography[129]. Complex dynamical phenomena manifest across a wide range of scales in the universe, from stellar systems to galaxies and even cosmological structures, and many of these can be effectively modeled using the tools of nonlinear dynamics. In a complementary vein, the Israeli physicist Oded Regev provides a comprehensive treatment of the subject in his book *Chaos and Complexity in Astrophysics*[130]. This work introduces the fundamental concepts of nonlinear dynamical systems, chaos theory, pattern formation, and complexity, and

---

[128]G. Contopoulos. *Order and Chaos in Dynamical Astronomy* (Springer, Berlin, 2002).
[129]G. Contopoulos. *Adventures in Order and Chaos: A Scientific Autobiography* (Kluwer, 2004).
[130]O. Regev. *Chaos and Complexity in Astrophysics* (Cambridge University Press, 2006).

illustrates their application to a variety of astrophysical problems. Collectively, these contributions underscore the significance of nonlinear science in modern astrophysics and highlight the field's ongoing vitality and interdisciplinary relevance.

4.3.2 *Chaotic rotation movement of moons of the solar system*

Discovered in 1848 in orbit approximately 1.5 million kilometers from Saturn, *Hyperion* is one of the largest irregularly shaped moons in the solar system. It has long been known that *Hyperion* exhibits chaotic rotational behavior, a consequence of both its highly elliptical orbit and irregular shape. The groundbreaking work of Jack Wisdom and collaborators[131] predicted this chaotic rotation through computational modeling.

Chaotic rotation about its center of mass implies that *Hyperion's* orientation and rotational velocity change in an unpredictable manner over time, rendering any long-term forecasting of its spin dynamics effectively impossible. In June 2005, NASA's *Cassini* spacecraft approached Hyperion and captured a sequence of images that, for the first time, directly revealed its chaotic rotational behavior. A decade later, in 2015, NASA's *New Horizons* mission discovered that two of Pluto's moons, *Nix* and *Hydra*, also exhibit chaotic rotation, further demonstrating the prevalence of such dynamics among irregular satellites.

---

[131] J. Wisdom, S.J. Peale, and F. Mignard. The Chaotic Rotation of Hyperion. Icarus 58, 137–152 (1984).

4.3.3 *Dynamic chaos in planetary systems*

The profound influence of nonlinear dynamics, particularly in the context of Hamiltonian systems, has been emphasized throughout this text. Russian astrophysicist I. I. Shevchenko offers a significant contribution to this field in his pioneering monograph *Dynamical Chaos in Planetary Systems*[132], which focuses on the stability and chaotic behavior of planetary systems and their subsystems. This work includes an exhaustive description of the main investigations that employ concepts of the resonant and chaotic dynamics of Hamiltonian systems, applied to both the dynamics of the solar system and the increasingly rich domain of exoplanetary systems. An essay review on the dynamics of planetary systems[133], which surveys several recent books, provides a broad perspective on the field and emphasizes key aspects also highlighted in the present discussion.

4.3.4 *Chaotic model of the universe*

A recent study by E. Aydiner[134] investigates the nonlinear interactions among dark energy, dark matter, and radiation within the framework of Friedmann–Robertson–Walker spacetime. The study introduces a simplified interaction model based on the time-dependent densities of these cosmic components and reveals the presence of a strange attractor in their dynamical behavior. This finding suggests that the temporal evolution of the universe may exhibit

---

[132]I.I. Shevchenko. *Dynamical Chaos in Planetary Systems* (Springer, Cham, 2021).

[133]M.A.F. Sanjuán. Dynamics of Planetary Systems. Contemporary Physics, 64(4), 315–318 (2023).

[134]E. Aydiner. Chaotic Universe Model. Scientific Reports 8, 721 (2018).

chaotic characteristics. Such an approach provides a novel perspective on cosmological evolution and opens new avenues for understanding large-scale structure and organization in the universe.

4.3.5 *Strange non-chaotic stars*

The unprecedented light curves of the Kepler space telescope have shown how the luminosity of some stars emits pulses at primary and secondary frequencies whose proportions approach the golden number, which is the most irrational number. A nonlinear dynamical system perturbed by an irrational proportion of frequencies generically exhibits a strange but non-chaotic attractor. The authors present evidence of the first observation[135] of strange non-chaotic dynamics in nature outside the laboratory. This discovery could contribute to the classification and detailed modeling of variable stars.

4.3.6 *Chaotic oscillations of neutrinos*

Neutrinos, extremely light and elusive particles, can play a crucial role in extreme astrophysical environments. One of their most distinctive properties is the ability to oscillate between different flavors as they propagate, a consequence of the non-alignment between their flavor and mass eigenstates. This phenomenon gives rise to a variety of rich and complex oscillation behaviors. A recent doctoral dissertation[136] and its associated publications, conducted at Aarhus

---

[135] J.F. Lindner, V. Kohar, B. Kia, M. Hippke, J.G. Learned, and W.L. Ditto. Strange Non-chaotic Stars. Phys. Rev. Lett. 114, 054101 (2015).

[136] Rasmus Sloth Hansen. Neutrino Oscillations in Very Dense Media: Production and Chaos. Ph.D. Dissertation, Aarhus University, Denmark, July 2015.

University in Denmark, investigate the emergence of chaotic phenomena in models of neutrino oscillations. These studies explore such dynamics in both the early universe and core-collapse supernovae, offering new insights into the nonlinear behavior of neutrino systems under extremely dense conditions.

4.3.7 *Chaos and complexity at the nano scale*

Efficient magnetization switching is a key requirement in both magnetic data storage and neuromorphic computing technologies. A recent experimental study has demonstrated that magnetic chaos[137], induced by an alternating torque, significantly enhances the magnetic switching rate in nanoscale ferromagnets. The study identifies a well-defined threshold in the torque amplitude and observes that switching efficiency is notably improved at lower frequencies. Through both experimental analysis and theoretical modeling, the authors quantitatively elucidate these findings, underscoring the critical role of low-dimensional magnetic chaos in facilitating efficient switching. This research highlights a compelling interplay between between chaos and stochasticity, paving the way toward improved energy efficiency in magnetic systems at the nanoscale.

4.4 Chaos and Complexity in Chemistry

Since the beginning of chaos theory, researchers have sought to apply its principles to the quantification of material structures and

---

[137]E.A. Montoya, S. Perna, Y.J. Chen et al. Reversal magnetization driven by low dimensional chaos in a nanoscale ferromagnet. Nat Commun 10, 543 (2019).

complex data, employing concepts such as strange attractors and fractal dimensions. Chaos theory offers powerful models for the in-depth analysis of complexity, making it a valuable tool across scientific disciplines. In their review article *Chaos Theory in Chemistry and Chemometrics*[138], Cramer and Booksh examine a wide range of experimental research in chemistry, focusing on the development of data analysis techniques grounded in chaos theory.

Many chemists have explored nonlinear and emergent behaviors through oscillating chemical reactions, such as the well-known Belousov–Zhabotinsky reaction. Bruce C. Gibb[139] argues that a deeper engagement with chaos and complexity may help address fundamental scientific questions, noting that "by embracing chaos and complexity, chemists can advance towards their goals, since they are well positioned to explain the complexities of life from the bottom up."

The importance of complexity in chemistry has been increasingly recognized over the past few decades[140]. As in many other scientific fields, chemistry has historically favored simplifying complex nonlinear phenomena through linear approximations. However, the study of complexity is emerging as a fruitful framework for addressing a wide array of challenges, particularly those related to the understanding of biological systems and the origins of life.

---

[138] J.A. Cramer and K.S. Booksh. Chaos theory in chemistry and chemometrics: a review. J. Chemometrics 20, 447–454 (2006).

[139] B. Gibb. Teetering towards chaos and complexity. Nature Chem 1, 17–18 (2009).

[140] G.M. Whitesides, R.F. Ismagilov. Complexity in Chemistry. Science 284 (5411), 89-92 (1999).

4.4.1 *Complex molecular systems and fractal nanomolecules*

Chemistry increasingly engages with complex molecular systems characterized by supramolecular interactions and intricate reaction networks. While the inherent complexity of these systems presents significant opportunities, such as the emergence of novel functions, progress has historically been limited by the lack of analytical tools capable of addressing such intricacy. In recent decades, however, growing interest in the concept of complexity has enabled the design of dynamical chemical networks and systems exhibiting emergent properties.

A recent article by Solà, Jimeno, and Alfonso[141] highlights examples of complex systems that support molecular recognition and catalysis. The authors advocate for a broader integration of complexity as a fundamental design principle in chemical research.

Additionally, it is well established that many natural structures exhibit fractal geometry, snowflakes and the branching patterns of leaves being common examples. Inspired by such phenomena, chemists have begun synthesizing molecules at the nanoscale with fractal-like geometries. Notably, Newcome et al.[142] demonstrated the nanoassembly of a molecular fractal based on the Sierpiński hexagonal gasket, opening possibilities for the development of novel photoelectric devices, molecular batteries, and energy storage technologies. Many researchers continue to explore the search for

---

[141] J. Solà, C. Jimeno and I. Alfonso. Exploiting complexity to implement function in chemical systems. Chem. Commun. 56, 13273-13286 (2020).
[142] G.R. Newcome et al. Nanoassembly of a fractal polymer: a molecular "Sierpinski hexagonal gasket." Science 312(5781), 1782-5 (2006).

extended molecular fractals formed by the self-assembly of small molecule components. Shang et al.[143] have carried out experiments on the preparation of Sierpiński fractal triangles to overcome the challenge of creating extended molecular fractals, by designing aromatic bromine compounds, managing to assemble defect-free Sierpiński triangles on Ag(111) surfaces. Their work suggests that halogen-hydrogen bonding plays a key role in directing the self-assembly process, offering a promising strategy for fabricating ordered molecular architectures with fractal geometries.

4.5 Chaos and Complexity in Life Sciences

4.5.1 *Fractals and Biology*

The growing influence of complexity science within the biological sciences is of significant importance. Concepts derived from fractal geometry and nonlinear dynamics have assumed a central role in both the description and modeling of a wide range of biological phenomena across multiple spatial and temporal scales. These theoretical frameworks have proven instrumental not only in molecular and cellular biology but also in understanding the organization of ecological systems and broader environmental processes. A comprehensive qualitative and quantitative overview of these developments is provided in the article by Kenkel and Walker[144], two researchers in the field of botany. Their work illustrates how fractal and nonlinear

---

[143] J. Shang et al. Molecular assembling Sierpiński triangle fractals. Nature Chem 7, 389–393 (2015).

[144] N.C. Kenkel and D.J. Walker. Fractals in Biological Sciences. Coenoses 11, 77-100, (1996).

models can be applied to diverse biological systems, reinforcing the inherently multidisciplinary nature of complexity science and its capacity to bridge traditionally distinct areas of research.

4.5.2 *Plant Physiology*

Vegetation dynamics have long been studied from a deterministic perspective, leading to important concepts such as climax, equilibrium, and reversibility. In recent decades, the emergence of chaos theory has changed our view of natural laws. New concepts such as disequilibrium, heterogeneity, perturbation and irreversibility have become increasingly popular. Some ecological successions in the special framework of vegetation typology and conservation ecology have even been considered stochastic and therefore their outcome is a matter of chance. The conclusion of Guillaume Decocq's work[145] is that all sequences are, at least in part, deterministic. Recent methods of nonlinear dynamics and complexity theory developed in theoretical physics can and should be applied to the description and analysis of systems in plant physiology, as stated by the authors of the article[146] where they use nonlinear dynamics as a tool for modeling in plant physiology. This is especially useful for the interpretation of temporal and spatial series, addressing rhythms and processes related to pattern formation. By applying theoretical concepts to a model system of circadian rhythmicity in plant physiology, the acid

---

[145]G. Decocq. Determinism, Chaos and Stochasticity in Plant Community Successions: Consequences for Phytosociology and Conservation Ecology. In: Gafta, D., Akeroyd, J. (eds) Nature Conservation. Environmental Science and Engineering (Springer, Berlin, Heidelberg, 2006).
[146]M.-Th. Hütt, U. Lüttge, Nonlinear Dynamics as a Tool for Modeling in Plant Physiology. Plant Biology 4(3), 281-297 (2008).

metabolism of *Crassulaceae*, nonlinear dynamics proves to be a remarkable tool to reveal the internal mechanisms that operate in the self-organization process of a system.

4.5.3 *Chaos theory and evolution*

Surprising as it may seem, methods derived from nonlinear dynamics have also begun to influence research in evolutionary theory. Paleoecologist Keith Bennett has argued that the evolutionary process exhibits characteristics typical of nonlinear systems. In a thought-provoking article titled *The Chaos Theory of Evolution*[147], he asserts that "We still have much to learn about how life evolved, but we will not develop a complete understanding until we accept the complexity of the system." Bennett contends that evolution should be understood not as a linear or purely deterministic process, but as a complex, nonlinear phenomenon shaped by feedback loops, bifurcations, and sensitivity to initial conditions.

In a subsequent publication[148], he explores how the interpretation of fossil and molecular phylogenetic records, particularly in relation to Quaternary climate changes, can be significantly improved by incorporating models based on the nonlinear dynamics between genotype and phenotype, as well as between climate and environment. According to this perspective, evolution, as described in *The Origin of Species*, is inherently unpredictable, chaotic, though not random, and is an inevitable consequence of reproduction unfolding

---

[147]K. Bennett. The chaos theory of evolution. New Scientist 208 (2782). pp. 28-31 (2010).
[148]K.D. Bennett. Is the number of species on earth increasing or decreasing? Time, chaos and the origin of species. Palaeontology 56, 1305-1325 (2013).

over time. Bennett proposes that biodiversity be conceptualized as a continually branching system of lineages, where species represent the terminal points of these branches. This view emphasizes a persistent increase in biodiversity and helps account for discrepancies often observed between genetic and morphological data across both spatial and temporal scales.

4.5.4 *Genetic circuits*

Genetics also uses methods of nonlinear dynamics. Suzuki and collaborators in a recent study[149] analyze the motifs of gene regulatory circuits that play crucial roles in cellular functions. Theoretical studies often omit time delays, but their inclusion alters the dynamics, generating diverse behaviors such as periodic, quasi-periodic, weakly chaotic, strongly chaotic and intermittent dynamics. Simulations suggest that a single feedback loop contributes periodic dynamics, while elements with two positive/negative loops exhibit chaotic dynamics. Likewise, they discuss the possible role of chaotic behavior in cellular robustness, such as in the context of cancer.

4.5.5 *Cell biology*

In another study[150], in this case in the context of cell biology, it is highlighted that "The importance of oscillations and deterministic

---

[149]Y. Suzuki et al. Periodic, Quasi-periodic and Chaotic Dynamics in Simple Gene Elements with Time Delays. Sci. Rep. 6, 21037 (2016).
[150]J. Werner, T. Pietsch, F.M. Hilker, H. Arndt. Intrinsic nonlinear dynamics drive single-species systems. Proc Natl Acad Sci US A. 119(44), e2209601119 (2022).

chaos in natural biological systems has been discussed for several decades and was originally based on discrete-time population growth models (May 1974). Recently, all types of nonlinear dynamics were shown for experimental communities where several species interact. Yet, there are no data exhibiting the whole range of nonlinear dynamics for single-species systems without trophic interactions. Up until now, ecological experiments and models ignored the intracellular dimension, which includes multiple nonlinear processes even within one cell type. Here, we show that dynamics of single-species systems of protists in continuous experimental chemostat systems and corresponding continuous-time models reveal typical characteristics of nonlinear dynamics and even deterministic chaos, a very rare discovery. An automatic cell registration enabled a continuous and undisturbed analysis of dynamic behavior with a high temporal resolution. Our simple and general model considering the cell cycle exhibits a remarkable spectrum of dynamic behavior. Chaos-like dynamics were shown in continuous single-species populations in experimental and modeling data on the level of a single type of cells without any external forcing. This study demonstrates how complex processes occurring in single cells influence dynamics on the population level. Nonlinearity should be considered as an important phenomenon in cell biology and single-species dynamics and also, for the maintenance of high biodiversity in nature, a prerequisite for nature conservation."

4.5.6 Neurosciences

The brain is widely regarded as one of the most complex systems known, and its study is pursued across numerous scientific disciplines, each offering a distinct perspective. Neurophysiology, neuropsychiatry, and the broader field of neuroscience have long sought to understand brain function, while computational neuroscience has emerged as a powerful interdisciplinary approach. For decades, physicists, mathematicians, and engineers have also shown considerable interest in modeling brain activity employing tools from nonlinear dynamics, chaos theory, nonlinear time series analysis, complex network theory, and the electronic modeling of neurons. Because many biological and neural systems are organized as networks of interacting oscillatory units, their functional behavior often arises from emergent collective dynamics. Among these, the synchronization of oscillations is a particularly prominent example, playing a crucial role in both healthy brain function and pathological conditions. In this context, Bick and collaborators[151] review recent methodological advances that apply nonlinear dynamical techniques to the study of neural oscillator networks. They emphasize how exact mean-field reductions can bridge the gap between reduced mathematical models and experimental data, thereby informing the development of novel therapeutic strategies for neurological diseases.

---

[151]C. Bick, M. Goodfellow, C.R. Laing, E.A. Martens. Understanding the dynamics of biological and neural oscillator networks through exact mean-field reductions: a review. J. Math. Neurosci. 10 (1), 9 (2020).

4.5.7 *Epilepsy*

Epilepsy is another field in which the application of nonlinear dynamics, chaos theory, and complexity science has seen significant development. In their seminal 2003 monograph *Epilepsy as a Dynamic Disease*[152], Milton and Jung conceptualize epilepsy as a dynamical disease. They advocate for interdisciplinary collaboration among clinicians, mathematicians, and bioengineers to better understand and model the mechanisms responsible for epileptic seizures, with the goal of improving control and treatment strategies. More recently, a review article by Moraes et al.[153] calls for a paradigm shift in epileptology by emphasizing the need to frame epilepsy as a dynamical system. This perspective challenges traditional linear models and opens new avenues for research and clinical intervention. Building on this evolving framework, a 2023 monograph by Scott and Mahoney[154] further advances the field by explicitly adopting a complex systems approach. The authors integrate theoretical and practical methodologies derived from complexity science to address the multifaceted nature of epilepsy, from seizure dynamics to therapeutic interventions.

---

[152]J. Milton and P. Jung, *Epilepsy as a Dynamic Disease* (Springer, Berlin, 2003).

[153]M.F.D. Moraes et al., Epilepsy as a dynamical system, a most needed paradigm shift in epileptology. Epilepsy Behav. 121, 106838 (2021).

[154]R.C. Scott, J.M. Mahoney. *A Complex Systems Approach to Epilepsy: Concept, Practice, and Therapy* (Cambridge University Press, 2023).

4.5.8 *Synthesizing life*

In recent years, the fields of Synthetic and Systems Biology have experienced remarkable growth, to the extent that they now constitute a distinct and rapidly evolving area of research. Jim Collins, originally recognized for his application of nonlinear dynamics and chaos theory to neuronal systems, made significant strides in modeling genetic switches using nonlinear differential equations, revealing the presence of bistable dynamics. His pioneering work played a foundational role in shaping synthetic biology, an inherently interdisciplinary field that, in its formative years, integrated concepts from nonlinear dynamics, complex systems physics, engineering, and molecular biology.

Among the landmark achievements in this domain are the construction of the *genetic toggle switch*[155] by Collins and collaborators, and the development of the *repressilator*[156], a synthetic genetic oscillator designed by Michael Elowitz and Stanislav Leibler. Both studies, published in Nature in 2000, represent foundational milestones in synthetic biology, demonstrating how engineering principles and nonlinear modeling can be harnessed to design and construct functional genetic circuits.

---

[155]T. Gardner, C. Cantor, and J. Collins. Construction of a Genetic Toggle Switch in Escherichia coli. Nature 403, 339–342 (2000).
[156]M.B. Elowitz, and S. Leibler. A Synthetic Oscillatory Network of Transcriptional Regulators. Nature 403, 335–338 (2000).

4.5.9   *Nonlinear dynamics and the heart*

Cardiac arrhythmias constitute disorders in cardiac rhythms and although in many cases the consequences are not serious, in others they can be fatal. In recent years, techniques from nonlinear dynamics have been increasingly applied to the analysis of electrocardiograms (ECGs) and cardiac rhythms, with the aim of improving both the understanding and control of arrhythmic behavior. An insightful overview of these approaches is provided by Karma and Gilmour[157], who present a broad survey of how nonlinear dynamics contributes to elucidating the phenomenology of cardiac arrhythmias, as well as to their prevention and potential control.

A more recent and comprehensive treatment is offered in a 2022 review article by Wouter-Jan Rappel, published in Physics Reports[158]. This work provides a quantitative exploration of cardiac arrhythmias through the use of nonlinear dynamics techniques and both two- and three-dimensional computational models. Emphasizing tools from nonlinear dynamics and statistical physics, the review highlights the profound insights these methods offer into the complex spatiotemporal dynamics underlying cardiac rhythm disorders.

---

[157] A. Karma, and R.F. Gilmour Jr. Nonlinear dynamics of heart rhythm disorders. Physics Today 60(3), 51-57 (2007).
[158] Wouter-Jan Rappel, The physics of heart rhythm disorders, Physics Reports 978, 1-45 (2022).

4.5.10 *Ecology*

Ecology is one of the fields within biology where dynamical systems theory has found extensive application, largely following the pioneering work of Australian physicist Sir Robert May, widely regarded as a founder of theoretical ecology. A compelling illustration of this approach is found in the book by ecologist Marten Scheffer, *Critical Transitions in Nature and Society*[159], which explores fundamental concepts such as oscillations and chaos, the emergence of patterns in complex systems, and adaptive responses in diverse contexts including lake dynamics, climate, evolution, and ecosystems. Scheffer employs mathematical tools such as bifurcation theory and long transient analysis to investigate ecosystem stability and critical phenomena such as desertification. For instance, a lake can be modeled as a nonlinear dynamical system, whose behavior may be governed by chaotic oscillations and complex fluctuations. Remarkably, this conceptualization dates back to 1887, when American entomologist Stephen Forbes, in his seminal essay *The Lake as a Microcosm*[160], proposed viewing lakes as self-contained ecological systems with intricate internal dynamics. Forbes, who has been recognized by the U.S. National Academy of Sciences as the founder of ecology in the United States, anticipated ideas that would later

---

[159]Marten Scheffer. *Critical Transitions in Nature and Society. Princeton Studies in Complexity* (Princeton University Press, 2009).

[160]Stephen A. Forbes. The Lake as a Microcosm. Bulletin of the Scientific Association (Peoria, IL), pp. 77–87 (1887).

be formalized through the lens of nonlinear dynamics and complex systems theory.

4.5.11 *Relevance of chaos in ecology*

In a recent article, Munch and collaborators[161] critically reexamine the prevalence and ecological significance of chaos. They argue that earlier skepticism, dating back to the 1970s, regarding the existence and relevance of chaotic dynamics in natural populations was largely driven by the technical difficulties associated with detecting chaos, particularly given the limitations of data resolution and the simplicity of population models available at the time. Contrasting this historical perspective, the authors point to a growing body of recent empirical and theoretical work that demonstrates the occurrence of chaotic dynamics in both laboratory and field-based ecological systems. They suggest that the previously perceived rarity of chaos may have been an artifact of methodological constraints rather than a reflection of biological reality. With advances in data collection, computational modeling, and nonlinear time series analysis, Munch and collaborators advocate for a renewed appreciation of chaos as a potentially common and ecologically meaningful feature of population and ecosystem dynamics.

4.5.12 *Ants, lions and chaos*

A striking recent example that has received considerable media attention involves the complex ecological interplay between ants and

---

[161]S.B. Munch, T.L. Rogers, B.J. Johnson, U. Bhat, C.-H. Tsai. Rethinking the Prevalence and Relevance of Chaos in Ecology. Annual Review of Ecology, Evolution, and Systematics 53, 227-249 (2022).

lions in the African savannah, an interaction deeply tied to the phenomenon of mutualisms in ecology. In a study published in Science[162], Kamaru and collaborators demonstrate how the invasion of a species of large-headed ants disrupted an existing ant-plant mutualism. This seemingly minor disturbance triggered a cascading ecological effect that ultimately altered predator-prey dynamics, with buffaloes replacing zebras as the primary prey of African lions. This case vividly illustrates the principle of sensitive dependence on initial conditions, a hallmark of chaos theory, highlighting how small perturbations in an ecosystem's structure can yield disproportionately large and far-reaching consequences. The study underscores the interconnectedness and fragility of ecological networks, reinforcing the relevance of nonlinear and complex systems approaches in understanding and managing biodiversity and ecosystem stability.

4.5.13   *How Life Works*

How Life Works: A User's Guide to the New Biology[163] is a recent and thought-provoking book written by Philip Ball, a British physicist and chemist who served as an editor at Nature for over two decades. In his book, he presents a transformative perspective on biology, highlighting the intricate interplay of complex systems that underpin living processes. Departing from traditional gene-centric views, he explores how life emerges from dynamic, multilevel interactions that defy simple, reductionist explanations.

---

[162]D.N. Kamaru et al. Disruption of an ant-plant mutualism shapes interactions between lions and their primary prey. Science 383, 433–438 (2024).
[163]Philip Ball. *How Life Works: A User's Guide to the New Biology* (Pan Macmillan, 2025).

Drawing from complexity theory, Ball emphasizes that "life is a hierarchical process, and each level has its own rules and principles: there are those that apply to genes, to proteins, to cells and tissues, and to body modules such as the immune and nervous systems." He underscores the multilevel, multidirectional, and hierarchical nature of biological organization, noting that "each level in the hierarchy of life's organization has its own rules, which are not sensitive to the fine details of those below." The book also touches on ideas of self-organization and dynamic landscapes, invoking metaphors such as "landscapes, basins, and channels" to elucidate how biological systems find stable yet adaptable forms.

A central theme is the concept of emergence across biological scales. He delves into how life expresses itself through the interactions of genes, proteins, cells, tissues, and organs, emphasizing that higher-level properties emerge from, rather than being strictly dictated by, lower-level components. This notion, often described as causal emergence, challenges the idea that causation resides solely at the molecular level. As he writes, "Research into complex systems has suggested how best to think more broadly about these operational principles of molecular and cell biology: the goal of these principles is to take causation largely out of the hands of the molecules themselves."

He also critiques the widespread use of mechanistic metaphors, such as the genome as a blueprint or cell as a machine, arguing that these oversimplify the flexibility, adaptability, and emergent behavior of biological systems. Instead, he advocates for conceptual tools that can better capture the richness and fluidity of life. Throughout the book, key ideas from dynamical systems theory, including noise and stochastic processes, show how these concepts help illuminate the probabilistic and often unpredictable nature of biological behavior.

Ultimately, interdisciplinary insights offer a compelling reevaluation of life itself, advocating a systems-level approach to biology that transcends reductionist paradigms and emphasizes context, interaction, and emergence as fundamental to understanding living systems.

4.6   Nonlinear Dynamics and Earth Sciences

Another major scientific domain in which the ideas of nonlinear dynamics have had significant impact is the Earth sciences. This broad field encompasses a range of disciplines, including atmospheric sciences, hydrology, and geology, all of which confront inherently complex and nonlinear systems. The application of nonlinear dynamics in these areas has been instrumental in addressing challenges related to prediction, particularly in the context of weather patterns, climate variability, seismic activity, and water resource management. Fractal and multifractal analysis techniques are widely used to characterize spatial and temporal patterns in geophysical data, while the identification and study of chaotic behavior have provided new insights into the limits of predictability in natural systems. Additionally, nonlinear time series analysis has become an essential tool for investigating complex signals arising from Earth processes, enabling researchers to uncover hidden structures, detect early warning signals, and improve forecasting capabilities.

Michael Ghil, an American mathematician and physicist, is renowned for his application of nonlinear dynamics to the study of climate systems and their interdisciplinary dimensions. In his comprehensive review article, *A Century of Nonlinearity in the Geosciences*[164], Ghil offers an overview of the development and impact of nonlinear ideas in mathematics and physics as applied to the Earth sciences over the past century.

Focusing primarily on mathematical concepts and methodologies, the article presents illustrative examples of how these tools have been, and continue to be, applied to both the solid Earth (i.e., the crust, mantle, and core) and its fluid envelopes, including the atmosphere and oceans. Ghil notes that terms such as nonlinearity, chaos, complexity, fractals, networks, tipping points, and turbulence have become key concepts in modern science, reflecting a paradigmatic shift in how natural systems are understood.

Published in celebration of the centennial of the American Geophysical Union, the article underscores the transformative role that nonlinear concepts and methods have played in advancing the geosciences. It also emphasizes the continuing and future relevance of these approaches in addressing the complex challenges posed by Earth system dynamics.

---

[164]M. Ghil. A century of nonlinearity in the geosciences. Earth and Space Science, 6, 1007–1042 (2019).

So-called slow earthquakes, like regular earthquakes, arise from unstable frictional sliding, which releases energy over a period of hours to months, rather than the seconds to minutes characteristic of a regular earthquake. In a recent work[165], the prediction capacity has been evaluated by studying the slip history in the Cascadia subduction[166] between 2007 and 2017. Their findings revealed that the system exhibits low-dimensional chaotic nonlinear dynamics rather than purely stochastic behavior. By reconstructing the underlying attractor, the authors estimated a finite predictability horizon on the order of days to weeks. These results suggest that, while short-term forecasting of slow earthquakes may be feasible, long-term prediction remains inherently limited, mirroring the challenges faced in predicting regular earthquakes, whose forecastability is similarly constrained to timescales comparable to their durations.

The American geologist Donald L. Turcotte has been a leading figure in advancing the application of fractal geometry and chaos theory within the fields of geology and geophysics. He is the author of the influential monograph *Fractals and Chaos in Geology and Geophysics*[167], which synthesizes concepts such as fractals, multifractals, chaotic dynamics, and self-similar time series to analyze and quantify complex natural phenomena, particularly those associated with Earth system processes. Turcotte's work has demonstrated the power of nonlinear and fractal-based approaches across a wide range

---

[165]A. Gualandi et al. The predictable chaos of slow earthquakes. Sci. Adv. 6, eaaz5548 (2020).

[166]The Cascadia subduction zone is a 700-mile fault that extends from northern California to British Columbia and is about 70-100 miles from the Pacific coast.

[167]Donald L. Turcotte. *Fractals and Chaos in Geology and Geophysics, 2nd ed.* (Cambridge University Press, Cambridge, 1997).

of geological disciplines, including seismology, mineralogy, geomorphology, structural geology, and geodynamics. These frameworks have been applied to quantitatively characterize phenomena such as drainage and erosion networks, flooding events, earthquake distributions, mineral and petroleum resource localization, rock fragmentation, mantle convection, and the generation of geomagnetic fields. His contributions have been instrumental in establishing a quantitative paradigm for understanding the inherent complexity and self-organization of geological processes.

Low-dimensional dynamical system models have also been employed to investigate geomagnetic reversals, phenomena in which the Earth's magnetic field undergoes polarity changes, resulting in the exchange of positions between the magnetic north and south poles. These reversals are well-documented in the geological record, particularly in magnetized rock formations, and serve as critical evidence in the field of paleomagnetism. Early models, such as the Rikitake two-disk dynamo system[168], revealed that chaotic behavior could emerge from simple deterministic systems and could plausibly account for the irregular timing of geomagnetic reversals. More recent work, such as that of Gissinger[169], has continued this line of research by developing refined deterministic models that reproduce key features of reversal dynamics within a chaotic framework. The study of geomagnetic reversals through low-dimensional chaos

---

[168]T. Rikitake. Oscillations of a System of Disk Dynamos. Proceedings of the Cambridge Philosophical Society 54, 89–105 (1958); Keisuke Ito. Chaos in the Rikitake Two-Disc Dynamo System. Earth and Planetary Science Letters 51, 451–456 (1980).
[169]C. Gissinger. A New Deterministic Model for Chaotic Reversals. Eur Phys J B 85, 1-12 (2012).

models has significantly advanced our understanding of the geodynamo, the mechanism by which Earth's magnetic field is generated, and has contributed to broader insights into the nonlinear nature of geophysical processes operating over geological timescales.

4.7 Applications in Physiology, Medicine and Biomedical Engineering

One of the earliest monographs dedicated to the exploration of physiological rhythms is *From Clocks to Chaos: The Rhythms of Life*[170], authored by physicists Leon Glass and Michael C. Mackey. This seminal work underscores the fundamental role of physiological rhythms in sustaining life, rhythms that, in many cases, persist throughout an organism's lifetime, where even brief interruptions can result in life-threatening conditions. Deviations of these rhythms beyond normal ranges, or the emergence of aberrant rhythms in systems that were previously stable, are frequently associated with disease states.

The monograph introduces key mathematical concepts, such as steady states, oscillations, and chaos, and applies them to the modeling and analysis of physiological systems. Many biological rhythms and physiological signals exhibit fluctuations that may be either stochastic or deterministic in origin, including chaotic dynamics. By employing mathematical models of biological oscillators, the authors explore transitions between dynamic regimes as well as the effects of external periodic forcing on system behavior.

---

[170]Leon Glass and Michael C. Mackey. *From Clocks to Chaos: The Rhythms of Life* (Princeton University Press, 1988).

Importantly, Glass and Mackey were among the first to propose the classification of certain pathological conditions as *dynamical diseases*, disorders arising from abnormal temporal organization in physiological systems. This concept has since influenced a wide range of interdisciplinary studies at the intersection of nonlinear dynamics and biomedical science.

In the field of biomedical engineering, A. Korolj and collaborators have recently conducted a comprehensive review[171] on the role of fractal analysis in quantifying chaos within biological systems, including the brain, muscles, eyes, lungs, and others, where such dynamics are closely linked to physiological function. The authors emphasize that optimal levels of chaos and fractality are often indicative of health in natural systems, whereas deviations from these levels may be associated with dysfunction or disease.

The review underscores the significance of maintaining a "healthy dose of chaos" in biomedical contexts and advocates for the integration of fractal principles into the design of materials, devices, and systems. By aligning engineering strategies with the intrinsic complexity and variability of living systems, the authors argue, it is possible to develop higher-fidelity biomedical technologies. Their work calls for the broader adoption of fractal frameworks in biomedical engineering research and development, highlighting the potential of these approaches to enhance both diagnostic tools and therapeutic devices.

---

[171]A. Korolj, H.T. Wu, M. Radisic. A healthy dose of chaos: Using fractal frameworks for engineering higher-fidelity biomedical systems. Biomaterials 219, 119363 (2019).

A recent milestone in the field of systems biology and biomedical modeling is the publication of *Systems Medicine: Physiological Circuits and the Dynamics of Disease*[172] by Israeli physicist Uri Alon, a pioneer in the development of systems biology. In this work, Alon lays the conceptual foundations for systems medicine, an emerging interdisciplinary approach aimed at addressing a fundamental question: Why do we develop certain diseases? Drawing on basic principles and mathematical reasoning, the book explores how physiological circuits, such as those governing hormonal regulation, immune responses, and aging, are structured, and how their inherent fragilities can give rise to specific pathological conditions. Through the use of simple nonlinear mathematical models, Alon illustrates how these circuits function under normal conditions and how their disruption can lead to disease. The work culminates in the proposal of a periodic table of diseases, offering a unifying framework that categorizes diseases based on shared circuit-level dynamics. This approach not only enhances our understanding of disease mechanisms but also suggests novel strategies for therapeutic intervention grounded in the logic of physiological design.

---

[172]Uri Alon. *Systems Medicine: Physiological Circuits and the Dynamics of Disease* (Chapman and Hall/CRC, New York, 2023).

Chapter 5

Perspectives. Complexity and Interdisciplinarity

Science is the belief in the ignorance of the experts[173].

— Richard P. Feynman

5.1 Introduction

Gerd Binnig (Fig. 5.1) is a German physicist renowned for co-inventing the scanning tunneling microscope (STM) alongside his Swiss colleague Heinrich Rohrer. This groundbreaking instrument enabled scientists, for the first time, to visualize and even manipulate individual atoms, revolutionizing the field of surface science and nanotechnology. For this seminal achievement, Binnig and Rohrer were jointly awarded the 1986 Nobel Prize in Physics. A few years later he wrote an essay *Aus dem Nichts*[174], dedicated to the creativity of nature and the human being, and where he reflects on nonlinear problems of physics, on the creativity of scientific activity and nature, fractal geometries, complexity, research on chaos, among other things.

---

[173]Richard P. Feynman. What Is Science. The Physics Teacher 7, 313–320 (1969).
[174]Gerd Binnig. *Aus dem Nichts: Über die Kreativität von Natur und Mensch.* (München: Piper Verlag, 1989). (From nowhere. About the creativity of nature and human beings.).

Fig. 5.1   Gerd Binnig. Physics Nobel Prize 1986.

Going so far as to make statements such as "I am convinced that the investigation of chaos will produce a revolution in the natural sciences similar to that which has taken place in quantum mechanics." Or this one: "Even scientists are unaware of most of what happens in science, unless one works in the same field in which an advance is made. An example of this is chaos research, an extremely interesting and fascinating field that I was completely unaware of. Now, after having become a little more familiar with it, I would dare to say that it is the most interesting field of research that exists today," where it highlights the research carried out in the field of nonlinear dynamics, the theory of chaos and complexity.

In this context of interdisciplinarity, it is worth recalling a brief historical note on the founding of the Academy of Mathematics of Madrid, established by King Philip II in 1582. A recent edition of *Institution of the Royal Mathematical Academy by Juan de Herrera*[175], originally published in 1584, includes a study by Prof. Pedro García Barreno of the Royal Academy of Sciences of Spain. In this study, García Barreno discusses the widely held view that the creation of the Academy was heavily influenced by Lulism, a philosophical and methodological current inspired by the thought of Ramon Llull (1232-1316), a 13th-century philosopher, logician, and mystic from Majorca, who wrote in Catalan and pioneered a system of logical reasoning to explore faith, knowledge, and universal truth.

Llull's ideas had a profound impact on intellectual figures such as Dante Alighieri, Pico della Mirandola, Nicholas of Cusa, Giordano Bruno, Paracelsus, René Descartes, and Gottfried Wilhelm Leibniz. He is also frequently regarded as a forerunner of modern computing due to his visionary attempts to formalize knowledge and logic.

In another scholarly contribution, García Barreno further explores the historical and intellectual continuity between the Academy of Mathematics and the Royal Academy of Sciences of Spain[176], tracing

---

[175] Juan de Herrera. Institution of the Royal Mathematical Academy. Edition by Juan Antonio Yeves Andrés. Preliminary studies by José Simón Díaz, Luis Cervera Vera, and Pedro García Barreno. Madrid: Institute of Madrid Studies, 197–231 (2006). (in Spanish)

[176] Pedro García Barreno, "The Madrid Academy of Mathematics of Felipe II," in The Royal Academy of Sciences 1582–1995, Madrid: Royal Academy of Exact, Physical and Natural Sciences, 1995, pp. 9–185. (in Spanish)

their shared lineage. The so-called Lullian science sought the integration of knowledge into a unified and harmonious whole, anticipating the ideals of holistic and interdisciplinary inquiry that would come to characterize the spirit of such academies. Several scholars have noted the influence of Lulism not only on the founding of the Academy of Mathematics but also on the intellectual vision of its first director, the Spanish architect, mathematician and geometrician Juan de Herrera (1530-1597).

5.2   Emergence, Complexity, Fundamental

The terms *emergence, complexity,* and *fundamental* are deeply relevant in scientific discourse, yet they are not interpreted in the same way by all members of the scientific community. This lack of consensus was evident in a series of letters published in Physics Today in 1986 and 1987, which centered around the controversy surrounding the Superconducting Super Collider[177]. These discussions sparked important reflections on the meaning of what is considered fundamental in science[178], as well as on broader concepts such as emergence, chaos, and the funding of scientific research. Notably, this debate included contributions from prominent figures such as Prof. Pedro M. Echenique, academician of the Royal Academy of Sciences of Spain, and Nobel Laureates in Physics Sheldon Glashow and Philip Anderson. Their exchanges revealed contrasting philosophical

---

[177]Charles J. Hailey, Gordon R. Freeman, Pedro M. Echenique, Sheldon L. Glashow. "Superconducting Super Collider." Physics Today 39(12), 11–15 (1986).
[178]Robert D. Black, Philip W. Anderson, Sheldon L. Glashow. "'Fundamental' Distinctions." Physics Today 40(8), 90–91 (1987).

perspectives on the priorities and foundations of science, illustrating the complexity and interdisciplinary nature of these fundamental concepts.

In summarizing the exchange of letters, it is worth noting that Pedro M. Echenique begins by critiquing Sheldon L. Glashow's use of the term *fundamental*, arguing that its broad applicability across nearly all scientific disciplines, including molecular biology, renders it ambiguous and potentially misleading. As an example of a genuinely fundamental principle, Echenique highlights the second law of thermodynamics, noting its foundational status despite being derived statistically. The discussion also touches on the concept of emergent properties, features that, while reducible to more basic components, cannot be straightforwardly deduced from them. The phenomenon of chaos, observed across a wide range of physical and biological systems, is cited as another illustration of this complexity. In response, Sheldon L. Glashow underscores the need for robust and sustained funding for the entire American scientific enterprise, framing it as essential not only for scientific progress but for the long-term survival and competitiveness of the nation.

5.3 Dialogue between Disciplines

In recent years, there has been increasing emphasis on the value of dialogue between disciplines as a source of inspiration for both new research questions and innovative solutions to longstanding problems. In the study of complexity, such interdisciplinary exchange is not merely beneficial but essential. Despite the diversity of methodologies employed across fields, the study of complex systems inherently involves challenges that span all areas of scientific inquiry. However, there is still a lack of true dialogue between disciplines that is so necessary for the advancement of knowledge of complex systems in particular and of science in general. Strengthening this dialogue is crucial, not only for the progress of complexity science itself but for the broader development of scientific knowledge in an increasingly interconnected world.

5.4   Interdisciplinarity and Language

Many scientists have noted that one of the principal obstacles to genuine interdisciplinarity, intrinsic to the sciences of complexity, is the absence of a common language. In this respect, the challenge is analogous to linguistic diversity: just as fluency in multiple spoken languages allows for greater expressiveness and the capacity to translate, intellectual fluency across disciplines requires not only understanding other methodological "languages" but also being able to translate among them. Each scientific discipline is rooted in its own conceptual frameworks, methodologies, and traditions, making it difficult to accurately interpret and engage with problems from other fields. The best way to face interdisciplinary problems is to be anchored in a discipline, otherwise you may be in the absence of true references. Without such grounding, there is a risk of superficiality or lack of depth. At the same time, academic training should place greater emphasis on interdisciplinary dialogue, much like how language education fosters familiarity with diverse forms of expression. Developing the ability to think across disciplinary boundaries is increasingly vital in the context of complex systems. Physics is often regarded as the foundational science, as all natural entities, including living organisms, are composed of elementary particles whose interactions physics seeks to describe. The remarkable success of the reductionist approach in physics largely stems from its focus on isolated systems. However, in reality, neither physical nor biological systems are truly isolated. Biological systems are inherently open, and in such systems, the environment is as crucial as the governing laws. While physics offers the underlying framework, it often overlooks key features of emergence inherent in biological complexity. As a result, a hierarchical structure arises, where each level of organization is governed by its own phenomenological laws. These higher-level laws emerge from, but are not reducible to, the fundamental laws of physics. This helps to explain why biologists can conduct meaningful and sophisticated research without needing detailed knowledge of quantum field theory, the Standard Model of particle physics, or nuclear physics. The autonomy of biological laws at higher levels reflects the richness and specificity of emergent phenomena in complex systems.

5.5 The Importance of Interdisciplinarity

The growing importance of interdisciplinarity is reflected in the increasing attention it receives in contemporary scientific literature.

A recent issue of the prestigious journal Nature Physics[179] features a collection of articles that highlight successful collaborations between researchers from different scientific fields. These contributions showcase how significant scientific advances have been achieved through multidisciplinary efforts (Fig. 5.2). The central theme of the issue focuses on the interface between physics and biology, illustrating how interdisciplinary interactions can lead to new insights and discoveries. The articles also offer practical advice on how to foster effective collaboration across disciplines. Although the discussion is centered on physics–biology interactions, the principles and strategies outlined are broadly applicable to interdisciplinary efforts across all scientific domains.

Fig. 5.2   Interdisciplinarity.

---

[179]Editorial. "Lost and Found in Translation." Nature Physics 19, 1735 (2023).

Several important ideas deserve emphasis when considering inter-disciplinarity. Chief among them is the ability to effectively communicate scientific concepts to individuals from other fields, an essential skill for successful multidisciplinary research. Establishing dialogue between disciplines is inherently challenging due to differences in research culture, terminology, and methodological approaches. Overcoming these barriers is not solely the responsibility of individual researchers; research institutions also bear a significant share of this responsibility. Furthermore, institutional efforts must be supported by appropriate funding structures and governmental policies that actively promote interdisciplinary collaboration.

If the integration of diverse research cultures leads to better science, the same principle applies to international collaboration. Communication barriers, in particular, remain one of the most significant obstacles to effective interdisciplinary interaction. The same term may carry different meanings in different fields, while distinct terminologies may be used to describe identical concepts. Yet this "multilingualism" of science can also be seen as a strength. Since the structure of language influences how we think, having multiple conceptual frameworks for approaching a problem can lead to a broader set of potential solutions.

Multidisciplinary collaborations that successfully bridge cultural and linguistic divides have yielded significant scientific advances, particularly in fields such as biological physics. A striking example is the resolution of the long-standing enigma of allometric scaling in biology, which explores why metabolic rates scale with body mass to the 3/4 power. A landmark paper published in Science in 1997[180] presented a general model for this phenomenon. The research was the result of a successful interdisciplinary collaboration between biologists James Brown and Brian Enquist and theoretical physicist Geoffrey West, former president of the Santa Fe Institute.

Geoffrey West later provided a compelling narrative of this work and its broader implications in his book *Scale: The Universal Laws of Life and Death in Organisms, Cities, and Companies*[181]. Additional contributions by West and colleagues appeared in both physics and biology journals[182], offering a unifying framework for understanding biological organization from the genomic to the ecosystem level[183]. This case exemplifies how crossing disciplinary boundaries, despite the inherent communication challenges, can lead to transformative scientific breakthroughs.

Curiosity and openness to alternative approaches will undoubtedly continue to drive progress in interdisciplinary research. This evolving landscape offers valuable lessons for researchers working at the interface of different scientific domains. To successfully overcome the barriers that traditionally separate disciplines, it is essential to foster a culture of interdisciplinary collaboration. As highlighted

---

[180]G.B. West, J.H. Brown, and B.J. Enquist. A General Model for the Origin of Allometric Scaling Laws in Biology. Science 276, 122–126 (1997).

[181]G.B. West. *Scale: The Universal Laws of Life and Death in Organisms, Cities and Companies* (Weidenfeld & Nicolson, UK, 2017).

[182]G.B. West and J.H. Brown. Life's Universal Scaling Laws. Physics Today 57(9), 36 (2004).

[183]J.H. Brown, J.F. Gillooly, A.P. Allen, V.M. Savage, and G.B. West. Ecology 85(7), 1771–1789 (2004); G. B. West and J. H. Brown. The Origin of Allometric Scaling Laws in Biology from Genomes to Ecosystems: Towards a Quantitative Unifying Theory of Biological Structure and Organization. The Journal of Experimental Biology 208, 1575–1592 (2005).

in recent discussions on strategies for multidisciplinary research[184], such a culture is foundational to achieving meaningful integration of diverse perspectives and methodologies.

Nonetheless, a significant obstacle remains in the current structure of research funding and evaluation, which is still largely organized along disciplinary lines. This system often fails to recognize or adequately support cross-disciplinary initiatives, making it difficult for truly integrative projects to thrive. Despite these challenges, the benefits of cultivating an interdisciplinary research culture are increasingly evident, offering not only scientific innovation but also broader societal relevance and impact.

5.6 Breaking Disciplinary Boundaries

When discussing complexity or interdisciplinarity, one frequently encounters the challenge of addressing similar concepts expressed in different disciplinary languages. However, beyond this linguistic and methodological difficulty lies another, perhaps more deeply rooted, issue within scientific and academic communities: the entrenched compartmentalization of knowledge. We have become so accustomed to the rigid boundaries between disciplines that it often seems unacceptable for someone from outside a given field to engage meaningfully with its core ideas, as though such an act constituted a transgression of disciplinary identity.

---

[184]T. Sanchis. Strategies for Multidisciplinary Research. Nature Physics 19, 1736–1737 (2023).

This defensiveness is also evident in the evaluation of inter-disciplinary research projects. Review and funding systems frequently compel applicants to categorize their work within a single disciplinary framework, undermining the very essence of interdisciplinary inquiry. While there have been some tentative and well-intentioned efforts, both nationally and internationally, to transcend these boundaries, they often remain limited in scope and ambition.

Greater institutional commitment is needed to foster a research culture that not only tolerates but actively encourages the breaking down of disciplinary boundaries. Doing so would not only enrich scientific inquiry but also better reflect the interconnected nature of the complex problems modern science seeks to address.

5.7 Complexity as an Opportunity

As the Hungarian physicist Tamas Vicsek insightfully observed in an essay published in Nature in 2002,[185] when a concept is not clearly defined, as is often the case with complexity, there is a risk of its overuse and misapplication. As a matter of fact, the term may be employed indiscriminately, sometimes merely as a marker of modernity, without a well-established theoretical foundation to support it. Despite this caution, his essay offers valuable insights into one of the central ideas underpinning complexity science: namely, that the laws governing the behavior of complex systems are qualitatively different from those that apply to their individual components. This perspective highlights a key departure from reductionist approaches,

---

[185]Tamas Vicsek. Complexity: The Bigger Picture. Nature 418, 131 (2002).

suggesting that emergent phenomena in complex systems require new conceptual frameworks and methods of analysis. It reinforces the notion that understanding the collective behavior of interacting elements often demands more than simply studying those elements in isolation.

5.8 Praise for Interdisciplinarity

As the academician Prof. Pedro R. García Barreno notes in his article *Cultural Integration: Transscience*[186], interdisciplinarity represents "one of the most inspired and fertile achievements; the most effective path, so far, towards knowledge." As both a training philosophy and a strategy for action, interdisciplinarity has yielded significant outcomes in areas ranging from environmental preservation and public health to scientific discovery, technological innovation, and deeper understanding of our place in the universe. Nevertheless, García Barreno also underscores the persistent challenges faced by those engaged in interdisciplinary research. Despite its evident benefits, researchers often encounter barriers that can lead to frustration or stagnation. These include difficulties in communication and collaboration across disciplines, what he refers to as "cultural" obstacles, as well as structural impediments stemming from the traditional organization of academic institutions. The entrenched division of universities, research organizations, professional societies, and

---

[186]Pedro R. García Barreno. Cultural Integration: Transscience. Revista de la Real Academia de Ciencias Exactas, Físicas y Naturales (España), Vol. 108, No. 1–2, pp. 1–12 (2015). (in Spanish)

scientific journals into disciplinary departments continues to limit the full realization of the interdisciplinary potential.

5.9 Simple Lessons on Complexity

The complexity of the world stands in stark contrast to the remarkable simplicity of the laws of physics. This paradox is one of the most striking aspects of the physical sciences: their governing laws are often elegant and succinct. Maxwell's equations for electromagnetism, Schrödinger's equation in quantum mechanics, and Newton's laws of motion in classical mechanics can each be expressed in just a few lines. As physicists Nigel Goldenfeld and Leo P. Kadanoff remarked in their essay *Simple Lessons from Complexity*, "Everything is simple and orderly — except, of course, the world."[187]

Indeed, when we observe the world around us, we are confronted with overwhelming complexity. While no universal "laws of complexity" currently exist in the same sense as the laws of physics, Goldenfeld and Kadanoff identify several simple lessons derived from the study of complex systems. These insights emerge from the analysis of diverse phenomena in the natural world. One foundational lesson is that nature is capable of generating intricate structures even under seemingly simple conditions, and conversely, that simple rules can sometimes emerge from highly complex systems. Another key idea is the importance of selecting the appropriate level of description in order to meaningfully capture the behavior of a given system.

---

[187]Nigel Goldenfeld and Leo P. Kadanoff. Simple Lessons from Complexity. Science 284, 87–89 (1999).

We are currently witnessing a flourishing of research in the science of complexity. However, it is important to recognize that this field demands a shift in perspective from the traditional mindset of physics. Rather than seeking universal laws that apply across all systems and contexts, complexity science acknowledges that each system may be unique. Yet, by studying one complex system, we can often extract transferable insights that help us understand others.

In this sense, physics, long grounded in reductionism and universality, may evolve toward a mode of inquiry more akin to the human experience: context-dependent, adaptive, and rich in interpretation. The science of complexity, then, not only expands the boundaries of physics but also reshapes our understanding of what it means to do science in a world that is as intricate as it is unpredictable.

5.10   Complexity as a Challenge Integrating Disciplines

Many of the core ideas and concepts associated with complexity pose significant challenges to the integration of knowledge across a wide range of disciplines. These include, among others, nonlinear dynamics and chaos theory, statistical physics, stochastic processes, information theory, network theory, engineering, the life sciences, and computer science. Naturally, this list is not exhaustive, but it illustrates the magnitude and scope of the intellectual challenge posed by complexity[188].

---

[188] Editorial. Complexity Matters. Nature Physics 18, 843 (2022).

Importantly, the goal of engaging with complexity extends beyond simply crossing disciplinary boundaries; it aspires to integrate diverse fields into a shared conceptual and methodological framework. A useful analogy is found in the parable of the blind men and the elephant (Fig. 5.3): each scientist may possess deep knowledge and an abundance of data concerning a specific part of the elephant, but without collaborative dialogue and interdisciplinary synthesis, no one can grasp the whole. The moral of the story underscores the urgent need for genuine interdisciplinarity. Isolated observations, no matter how sophisticated, are insufficient to comprehend the full nature of complex phenomena.

This parable is eloquently retold in the poem *The Blind Men and the Elephant* by the American poet John Godfrey Saxe (1816–1887), which can be found in the reference[189]. In this context, the story serves as a compelling metaphor for the fragmented nature of scientific inquiry when pursued in isolation, and the transformative potential of integration across disciplines.

5.11   Birds and Frogs

In his 2008 AMS Einstein Lecture, physicist and mathematician Freeman Dyson presented a memorable allegory in an essay titled *Birds and Frogs*[190], which, although written with mathematicians in mind, can be readily applied to physicists and scientists more

---

[189]J. Himmelfarb, P. Stenvinkel, T.A. Ikizler, R.M. Hakim. The elephant in uremia: Oxidant stress as a unifying concept of cardiovascular disease in uremia. Kidney International 62, 1524—1538 (2002).

[190]Freeman Dyson. Birds and Frogs. Notices of the American Mathematical Society 56(2), 212–223 (2008).

Fig. 5.3 The elephant and the six blind men. Cartoon taken from J. Himmelfarb, P. Stenvinkel, T.A. Ikizler, R.M. Hakim. The elephant in uremia: Oxidant stress as a unifying concept of cardiovascular disease in uremia. Kidney International 62, 1524–1538 (2002), originally property of the authors; G. Renee Guzlas, artist.

broadly. Dyson begins by defining his metaphor: "Some mathematicians are birds, others are frogs. Birds fly high in the air and survey broad vistas of mathematics out to the far horizon. They delight in concepts that unify our thinking and bring together diverse problems from different parts of the landscape. Frogs live in the mud below and see only the flowers that grow nearby. They delight in the details of particular objects, and they solve problems one at a time."

Dyson further elaborates by drawing a historical parallel between two philosophical archetypes: the Baconians, represented by Francis Bacon and classified as frogs, and the Cartesians, embodied by René Descartes and likened to birds. Throughout the history of science, progress has been made through the interplay and collaboration of these two cultures, those who work toward grand unifying theories and those who focus on solving concrete, localized problems.

In the context of nonlinear dynamics, chaos theory, and complexity science, this dual perspective is not only relevant but essential. Given the relatively young and rapidly evolving nature of these fields, the contributions of both "birds" and "frogs" are indispensable. On the one hand, detailed investigation into specific problems is necessary to address the many open questions and emerging challenges. On the other, the continued development of new methods and conceptual frameworks may lead to broader unification and deeper understanding. As the field matures, it is the dynamic collaboration between these two scientific temperaments that will drive its progress forward.

Chapter 6

Teaching Complexity and Nonlinearity

The most successful people are those who are good at plan B. For me,
research is not trying to solve a hard problem. Rather, it is taking
a question and an approach and continually changing both, gaining
understanding, until they fit together perfectly. I love to find ideas
that surprise, that can be explained simply, and that people need to
know. A degree in mathematics is a license to explore the universe[191].

— James A. Yorke

6.1   Why Teaching Complexity Matters

So far, we have explored the history and development of nonlinear dynamics, chaos and complexity, its diverse applications in the
scientific disciplines, and the epistemological implications that underscore the importance of interdisciplinarity. We have gained a
broad overview of the potentiality of the ideas related to the field, as
well as the relevance in the development of science itself. Yet, a key
pressing question remains: how is this knowledge taught, integrated,
and sustained within contemporary science education. This corpus

---

[191]James A. Yorke. The James A. Yorke Rotunda. University of Maryland. 2013. Fig. 6.1.

of knowledge, this new framework needs to be transmitted to the younger generations.

Fig. 6.1   The plaque containing the quotes from James A. Yorke is located in the James A. Yorke Rotunda at the Mathematics building of the University of Maryland, inaugurated in 2013.

Needless to say, the theme of interdisciplinarity and complexity continues to generate considerable interest within the scientific community. And surprisingly in all scientific disciplines. I recall a conversation I once had with a historian about the challenges of translating the ideal of interdisciplinarity into academic practice. It is, without doubt, a difficult endeavor. As I have noted in the previous chapter, it requires crossing disciplinary boundaries and implies making a genuine effort to understand the perspectives and methods of other scientific fields, an effort that is intellectually demanding and not always immediately rewarding. This is especially true in times when hyperspecialization has been so strongly emphasized. In spite of the ideal that many would embrace, it is also something not for everybody.

When asked whether a different dynamics is possible within academia and research, my response is unequivocally affirmative. I am fully aware of the obstacles involved, but I believe that fostering genuine interdisciplinary engagement is both possible and necessary. This is, of course, a delicate matter. On the one hand, no one can participate meaningfully in interdisciplinary dialogue without first having mastered at least one discipline. On the other, if we believe such dialogue is essential for addressing the complexity of the world, then change must begin somewhere. This point is very relevant, since we need to avoid superficiality. Many proverbs across cultures warn us of the idea that "he who grasps too much, holds onto little", or as we say in Spanish, "el que mucho abarca, poco aprieta". At the same time, if we are to pursue the goal of interdisciplinarity, we must begin acting on it without delay.

Precisely, in my view, that starting point must be during the university years. Interdisciplinary thinking should not be postponed until later stages of academic development. To this end, special undergraduate programs with a stronger interdisciplinary orientation should be implemented. There are already promising models in place at several European and American universities. Postgraduate and doctoral training could also offer valuable opportunities for interdisciplinary exposure, particularly when supported by well-designed curricula and faculty who are committed to the ideals of interdisciplinary education.

Of course, I do not pretend that this transformation will be easy. But I do not believe it is impossible, especially if there is a genuine will to make it happen. Crucially, it will require strong support and commitment from academic and scientific leadership. In fact, I am convinced that the need for interdisciplinary approaches will only grow in the future, and that they will become not just desirable, but inevitable. If we are to prepare the next generation of scientists and scholars for the complexity of the problems they will face, we must begin now, by rethinking how we train them, and by embracing a more open, integrative vision of knowledge itself.

Certainly, there is nowadays a growing global interest in education on complexity theory. Increasingly, institutions are rethinking rigid disciplinary frameworks by creating programs that introduce students to complex systems, nonlinear dynamics, and computational modeling. Concepts once confined to advanced research, like bifurcations, chaos, fractals, and network theory, are beginning to appear in mathematics, physics, chemistry, biology, geology, and computer science curricula, though needless to say, the processes are progressing very slowly. Fields like biomedical engineering[192] and environmental science[193], in particular, highlight the need for tools that can handle feedback, emergence, and nonlinear interactions across scales.

In a world shaped by rapid technological evolution, climate instability, and biospheric tipping points, understanding complexity is no longer optional, it is essential. Phenomena such as atmospheric dynamics, earthquake systems, protein folding (despite breakthroughs like AlphaFold)[194], metabolic networks, and climate modeling all involve nonlinear, multiscale behavior that cannot be reduced to isolated variables. Educating future scientists and engineers to work with such systems demands a shift from reductionist paradigms toward integrative, dynamic thinking. Yet, despite its relevance, complexity science remains largely underrepresented in traditional science and engineering education. As is well known, curricula often emphasize idealized, closed-system problems with exact solutions, leaving students unprepared for the open-ended, data-rich, and feedback-driven systems that dominate real-world challenges.

Because teaching complexity is not simply about adding new topics to a traditional curriculum, it is about cultivating a new scientific mindset. And this is what makes it more challenging and interesting. New ideas and methods appear to the students. They must

---

[192]R.D. Kamm, R. Bashir. Creating Living Cellular Machines. Annals of Biomedical Engineering 42(2), 445-259 (2014).

[193]S. Levin. *Fragile Dominion: Complexity and the Commons.* Perseus Books, 1999. D.C. Krakauer and M.A. Nowak. Book Review on Fragile Dominion: Complexity and the Commons. Notices of the AMS 47(5), 564-568 (2000).

[194]J. Jumper, R. Evans, A. Pritzel, et al. Highly accurate protein structure prediction with AlphaFold. Nature 596, 583–589 (2021).

learn to model dynamical systems, visualize evolving patterns, and understand stability, sensitivity, and emergence. This requires not only mathematical and computational fluency but also an intuitive grasp of nonlinear behaviors. Hands-on simulations, such as modeling fluid turbulence, chemical oscillations, neural networks, or population dynamics, can bridge abstract theory with tangible understanding. These experiences deepen analytical skill while nurturing systems thinking.

Complexity also fosters collaboration across the natural and computational sciences. It provides a shared language where physicists, mathematicians, chemists, biologists, geologists, and engineers can converge around questions that no single field can resolve alone.

6.2 Integrating Complexity into Science Education

Numerous problems appear in this endeavor. Effectively teaching complexity in the natural sciences faces significant hurdles rooted in the historical organization of academic disciplines. Over centuries, specialization has led to the compartmentalization of knowledge, physics taught apart from biology, chemistry isolated from geology, and engineering separate from mathematics. This has been the way we have been taught in previous generations, making us feel that every discipline has little or nothing to deal with the rest. While this has deepened disciplinary expertise, it has also hindered students from understanding integrated, nonlinear systems that characterize real-world scientific challenges. And this is the real problem.

As already mentioned, phenomena like climate systems, neural dynamics, protein folding, and geological processes involve interactions across scales and disciplines. However, traditional curricula often emphasize only linear models, closed systems, and problems with precise solutions, approaches that are poorly suited to capturing the feedback, emergence, and unpredictability at the heart of complex systems. During the years, I have been witness of excellent educators that they have made their best to change this in their courses. In particular, in the field of mathematics attempting to teach a dynamical systems view point instead of traditional courses on differential equations where emphasis had been mainly in learning techniques to obtain closed and exact solutions. Furthermore, this gap is further exacerbated by curricular rigidity, which limits opportunities for interdisciplinary integration and adaptive learning. While some university systems offer greater flexibility in curriculum design, many remain constrained by structures that are ill-equipped to address the complexity of real-world phenomena. Furthermore, standard science education often prioritizes measurable outcomes and well-defined problem sets. In contrast, complexity science thrives on open-ended inquiry, exploratory modeling, and tolerance for uncertainty, traits that conventional assessment systems find difficult to accommodate.

Another challenge lies in disciplinary language and methodology. This was already discussed in the previous chapter. The issue is that each field has developed its own specialized vocabulary and tools, valuable for depth but often obstructive to cross-disciplinary communication. Teaching complexity requires actively bridging these divides, helping students recognize and translate concepts such as equilibrium, networks, phase transitions, or chaos across contexts in physics, biology, and beyond. And this is also challenging.

Complexity also demands comfort with ambiguity and counter-intuitive thinking. Concepts like emergence or sensitivity to initial conditions challenge traditional assumptions in physics and engineering about predictability and control. For example, chaotic systems, demonstrated through models like the double pendulum or the Lorenz attractor, show that deterministic rules can yield unpredictable outcomes. Using simulations and hands-on models, such as chemical oscillators or ecological interactions, helps build the necessary intuition. Developing this systems-oriented, nonlinear perspective is essential for modern scientific literacy. Encouraging students to iterate, reflect, and explore complex behavior fosters intellectual adaptability and critical reasoning.

Finally, integrating complexity into natural science education requires more than adding interdisciplinary content, as it was previously noted. It calls for a structural shift in how science is taught, emphasizing dynamical systems, cross-disciplinary thinking, and the tools to navigate complexity. This prepares students not only for research, but for solving the deeply interconnected challenges of the 21st century.

In any case, there are many initiatives of academic institutions that have begun integrating complexity science into curricula through innovative interdisciplinary programs across the globe.

Notable examples include complex systems centers in physics departments, applied mathematics tracks in dynamical systems, and computational biology programs that train students to work across scales from molecules to ecosystems. These efforts highlight growing international recognition of the value of complexity education across natural and computational sciences, and provide valuable models for curricular design.

Among various institutional initiatives at different curricular levels, some examples of integrative courses also stand out, where interdisciplinarity and complexity play a significant role. One notable example is *The 2023 Kanpur Lectures Series on Engineering and Science in our World by Prof. Sandip Tiwari*[195]. Prof. Sandip Tiwari is the Charles N. Mellowes Emeritus Professor of Engineering at Cornell University and a Distinguished Visiting Professor at the Indian Institute of Technology at Kanpur. The contents of these lectures are available in a monograph titled *Engineering and science in our world. This I believe*[196].

Besides reflecting deeply on education, he addresses several real-world problems through an innovative lens, drawing from complexity theory to offer new insights. He frames this work as part of a broader educational initiative, affirming: "We must embrace complexity as a way to become leaders." Moreover, emphasizing the integration of diverse issues under an open and shared conceptual framework,

---

[195]The 2023 Kanpur Lectures Series on Engineering and Science in our World. `https://shorturl.at/zsgk8`.

[196]Sandip Tiwari. Engineering and science in our world. This I believe (2023). `https://shorturl.at/cnIz7`.

he writes: "Reality is open, we don't know all, we don't know even what all is, there are interactions of all different kinds with an environment that is not immutable. Even the universe is expanding. Complex problems arise in openness, large dimensionality, large number of interactions, spontaneous events, et cetera, and tackling them in probabilistic terms, that is, getting to a good-enough or more likely answer, is to give up the guarantees."

6.3 Teaching Tools for Complex Systems

Given the conceptual challenges posed by complexity and nonlinear dynamics, effective science education in this area requires tools that go beyond traditional lectures and textbooks. Fortunately, advances in computational technologies, interactive simulations, and visualization techniques have created powerful opportunities for teaching complex systems, particularly in the natural sciences and engineering.

One of the most impactful tools is computer simulation. Simulations allow students in fields like physics, biology, and biomedical engineering to explore dynamical systems that exhibit chaos, bifurcations, oscillations, or emergent behavior. Using software designed for dynamical systems, students can model chemical reactions, fluid turbulence, gene regulation networks, or planetary orbits, gaining direct experience with how small changes in parameters lead to qualitative shifts in behavior. These explorations build intuitive understanding alongside analytical skill.

Interactive visualizations further support conceptual learning. Visualizing phase space trajectories, strange attractors, or bifurcation diagrams helps students grasp abstract mathematical behaviors in a tangible form. For instance, animations of the Lorenz attractor or the logistic map can illustrate sensitive dependence on initial conditions, basic ideas in chaos theory that underpin real systems in meteorology, neuroscience, and population dynamics.

Hands-on activities rooted in scientific modeling can also be adapted across disciplines. Students might simulate chemical oscillators like the Belousov-Zhabotinsky reaction, model earthquake recurrence using spring-block systems, or explore predator-prey dynamics in ecosystems. These projects are particularly useful in fields like biology, environmental science, and biomedical engineering, where complex feedback and adaptation are key.

Although less technical, analogies and metaphors remain essential for translating abstract ideas into familiar terms. Storytelling, especially around historical discoveries or unexpected experimental results, can also contextualize concepts and sustain engagement. Consequently, teaching complexity in the natural sciences benefits from a combination of computational, visual, experimental, and narrative tools. Together, they offer students not just knowledge, but an active, intuitive grasp of how complex systems behave, evolve, and interact, preparing them to apply these insights across scientific domains.

6.4   Complexity in Undergraduate vs. Graduate Curricula

A key consideration in teaching complexity science is aligning curriculum design with the educational level of students. Undergraduate and graduate programs differ significantly in goals, depth, and methodological expectations. Therefore, complexity education should be tailored accordingly, especially within fields such as physics, biology, mathematics, chemistry, geology, computer science, and biomedical engineering.

At the undergraduate level, the primary objective is to build foundational understanding and intuition. Courses should introduce basic concepts such as nonlinear dynamics, deterministic chaos (e.g., the logistic map or double pendulum), basic network theory, and elementary computational modeling. Students in physics or engineering, for instance, might explore oscillatory systems and phase transitions, while those in biology or environmental science could examine predator-prey models or feedback loops in ecosystems. Emphasis should be placed on hands-on simulations, visualizations, and real-world case studies, ranging from chemical reactions to planetary systems, to help students grasp how complex patterns emerge from simple rules. Assessment at this stage should prioritize conceptual clarity, analytical reasoning, and the ability to interpret nonlinear phenomena intuitively.

An experience on a course for beginners is offered in *Teaching Nonlinear Dynamics and Chaos for Beginners*[197]. The article

---

[197] Jesús M. Seoane, Samuel Zambrano and Miguel A. F. Sanjuán. Teaching Nonlinear Dynamics and Chaos for Beginners. Lat. Am. J. Phys. Educ. 2(3), 205-211 (2008).

presents a decade-long experience implementing an elective under-graduate course on Nonlinear Dynamics, offered across a wide range of science and engineering degrees. Nevertheless, the course has continued to be imparted for students of an interdisciplinary degree in science at my university. The course attracted students from diverse backgrounds, particularly Chemical Engineering, Environmental Sciences, and Computer Science, highlighting the broad relevance and appeal of the subject. The central objective of the course has been to introduce chaotic phenomena in physical systems using only minimal prerequisites in physics and mathematics. Instead of emphasizing formal mathematical rigor, the course has focused on conceptual understanding through visual and interactive learning. A key pedagogical strategy has involved using numerical simulations by using different software materials, allowing the students to explore complex dynamical behavior without advanced mathematical tools.

Additionally, the course included a virtual nonlinear physics laboratory, allowing students to interact directly with experiments that illustrate nonlinear and chaotic behavior. This hands-on component reinforced theoretical material and helped cultivate a deeper, more intuitive understanding. Building on established teaching models, the course demonstrated that Nonlinear Dynamics can be made both accessible and intellectually engaging for undergraduates, regardless of their academic background or specialization.

At the graduate level, the curriculum must advance in both depth and technical sophistication. Students should explore complex adaptive systems, advanced network analysis, stochastic processes, agent-based modeling, and computational techniques such as machine learning applied to scientific data. Specifically, graduate students in computer science might simulate distributed systems or neural networks; those in biomedical engineering could model physiological networks or disease dynamics. Original research becomes central, whether through simulation studies, mathematical modeling, or experimental design, focusing on open-ended, interdisciplinary problems.

Additionally, graduate education should incorporate training in research methodology, scientific communication, and interdisciplinary collaboration. Literature review, critical analysis, and publication-quality research projects should be integral components. These experiences prepare students not only to engage with complexity theory but to contribute meaningfully to its development.

In summary, undergraduate complexity education builds broad, intuitive understanding, while graduate programs cultivate technical expertise and research independence, both essential for preparing scientists and engineers to confront the complexity of today's scientific challenges.

6.5 Interdisciplinary Education: Breaking Institutional Barriers

The essence of complexity science lies in its inherently interdisciplinary nature. However, traditional educational systems often separate knowledge into rigid disciplinary barriers. Complex phenomena, such as climate systems, tectonic processes, protein folding, neural activity, or atmospheric dynamics, cannot be fully understood from a single-disciplinary perspective. Teaching complexity requires integrating insights from physics, mathematics, biology, chemistry, geology, computer science, and engineering.

Conventional curricula emphasize depth within individual fields, but this often limits students' ability to recognize the interdependence between natural systems. In contrast, interdisciplinary education promotes synthesis, such as connecting differential equations in physics with biological feedback mechanisms, or using network theory to explore interactions in chemical or geological systems. This approach cultivates a broader and more holistic scientific understanding.

To be effective, interdisciplinary education must also address conceptual and methodological gaps across fields. Courses that explicitly draw connections between disciplines help students develop fluency in multiple scientific languages, whether that means understanding phase transitions in both materials science and atmospheric dynamics, or applying mathematical models to biological growth and chemical reactions. By encouraging students to embrace complexity and uncertainty, interdisciplinary teaching nurtures intellectual adaptability, critical thinking, and creative problem-solving, skills essential in research, engineering, and scientific innovation.

Finally, embedding complexity within interdisciplinary education equips future scientists and engineers to understand and engage with the multifaceted, interconnected nature of the natural world.

6.6 The Role of Narrative, Metaphors, and Historical Context

Complexity science benefits from teaching approaches that leverage narrative, metaphor, and historical context, especially in highly abstract or technical fields like mathematics, physics, and computational modeling. Narratives provide accessible entry points into difficult concepts. Stories about early discoveries, like Poincaré's work on celestial mechanics, Lorenz's atmospheric simulations that led to chaos theory, or Mandelbrot's development of fractals, humanize the subject and ground abstract ideas in real scientific exploration.

Metaphors function as cognitive bridges. Comparing bifurcations in dynamical systems to chemical phase changes or geological tipping points helps students connect complex concepts to familiar natural processes. Similarly, metaphors like the "butterfly effect" or "strange attractors" allow students to grasp sensitive dependence or long-term unpredictability in physical systems. Historical and metaphorical framing not only makes learning more engaging, but also reinforces understanding by connecting theoretical constructs to the real, evolving world of science.

6.7 Educating for Complexity: Building Curricula for a Nonlinear World

In the 21st century, understanding complexity has become a foundational component of scientific literacy. As global challenges, from climate instability and pandemics to metabolic networks and energy transitions, grow increasingly interconnected and nonlinear, the ability to think in terms of systems, feedback, and emergence is no longer optional. Preparing students to navigate and shape this reality requires thoughtful, future-oriented curricula grounded in complexity science.

A robust complexity curriculum begins with key conceptual foundations: nonlinear dynamics (chaos, bifurcations, fractals), network theory, agent-based modeling, stochastic processes, and information theory. These topics are highly relevant across natural science and engineering disciplines, including physics, biology, chemistry, mathematics, geology, computer science, and biomedical engineering.

Equally essential is computational fluency. Students must learn to write code, run simulations, and visualize dynamic data. These skills are now basic tools in scientific inquiry, from simulating biochemical pathways and weather systems to modeling tumor growth or neural activity.

However, theoretical knowledge alone is not enough. Hands-on, problem-driven learning is vital for building intuition. Engaging students in modeling real systems, like fluid turbulence, epidemic spread, or ecological interactions, helps bridge abstract theory with applied insight. Such experiential learning encourages exploration, iteration, and creative reasoning.

Assessment methods should reflect the interdisciplinary and integrative nature of complexity. In addition to traditional exams, evaluations should include computational projects, simulations, group work, and reflective writing to capture the depth and breadth of student understanding.

Interdisciplinary collaboration must be embedded in educational design. Lab-based courses, collaborative workshops, and mixed-discipline seminars provide spaces where students from diverse scientific fields learn to share methods and perspectives. This reflects how complexity operates in the real world, across boundaries, through interaction, and often in unexpected ways.

Among the various resources available for approaching complex and nonlinear systems, Klaus Mainzer's *Thinking in Complexity: The Computational Dynamics of Matter, Mind, and Mankind*[198] is particularly relevant for educational purposes, as it combines a broad interdisciplinary scope with a philosophical and historical perspective that complements more technical treatments of the subject.

Finally, institutional support is critical. Implementing complexity education requires flexible curricula, faculty development, and structures that encourage cross-departmental collaboration. Complexity should not remain a niche elective, it should become a central feature of contemporary science education. In the end, teaching complexity equips students not just to understand the nonlinear nature of the world, but to act within it, critically, creatively, and responsibly.

6.8   Complexity, Nonlinearity and AI

There are currently several issues perceived by the public as critically important, such as climate change, global warming, the management of new intermittent energy sources, biodiversity loss, crowd

---

[198]Klaus Mainzer. *Thinking in Complexity. The Computational Dynamics of Matter, Mind, and Mankind.* Springer-Verlag, Fifth Revised and Enlarged Edition, 2007.

dynamics, migration (of humans, animals, and plants, including invasive species), sustainable forest management, epidemic prevention, the emergence of new diseases, and the rise of cryptocurrencies and blockchain technologies. These are undoubtedly complex challenges that cannot be addressed adequately by a single discipline. Instead, they require an interdisciplinary approach and are rightly classified as complex problems.

Artificial intelligence (AI) must also be included in this discussion, as we increasingly face its influence across all levels of science and society. In a recent essay[199], a discussion was presented on the relationship between AI, chaos, prediction, and scientific understanding. A key issue explored was the role of understanding in science, especially in contrast to AI tools that excel in prediction but often lack interpretability.

We should therefore reflect on the connections between AI, nonlinearity, and complexity, and consider their implications for teaching. Otherwise, this chapter, which aims to promote a vision of teaching in the future, could seem outdated.

As an illustrative example, consider protein folding. The neural network-based model AlphaFold2 has successfully predicted the structures of over 200,000 proteins, as mentioned earlier[200]

This raises a fundamental question: how should a young biology student approach their career today? Should they focus on learning

---

[199]Miguel A.F. Sanjuán. Artificial Intelligence, Chaos, Prediction and Understanding in Science. *Int. J. Bifurc. Chaos Appl. Sci.* **31**(11), 2150173 (2021).

[200]J. Jumper, R. Evans, A. Pritzel, et al. Highly accurate protein structure prediction with AlphaFold. Nature 596, 583–589 (2021).

the underlying complexity of biological systems, or instead concentrate on mastering tools like AlphaFold2? Needless to say, dilemmas like this one, calls for deeper philosophical reflection.

On the one hand, AI offers powerful tools that can enhance the illustration and visualization of chaotic phenomena, contributing to better understanding. On the other hand, AI itself exemplifies complexity and nonlinearity. It operates through systems with inherently rich and unpredictable dynamics, which students already encounter in their daily lives.

6.9 Concluding Reflections

Teaching complexity is not merely a shift in content, but a shift in mindset. In a world shaped by interconnected natural systems and technological transformation, the ability to think across boundaries, navigate uncertainty, and recognize emergent patterns is essential.

Throughout this chapter, it has been argued that complexity education nurtures precisely these capacities. It challenges disciplinary constraints, promotes systems thinking, and equips students to confront nonlinear, real-world phenomena, whether modeling brain activity, analyzing fluid dynamics, or simulating planetary systems.

Barriers remain: institutional inertia, rigid curricula, and discipline-specific assessment. But the success of pioneering programs worldwide shows that change is not only possible, but necessary.

Teaching complexity aligns with the deeper goals of science education, developing not just specialists, but adaptable, integrative thinkers prepared to lead in uncertain and evolving contexts.

Moving forward, institutions must invest in interdisciplinary collaboration, flexible teaching models, and the integration of complexity into basic science curricula. Educators must encourage creativity, curiosity, and resilience in the face of ambiguity.

To teach complexity is to prepare future scientists, engineers, and thinkers not only to analyze the world, but to engage with it meaningfully. It is a call to cultivate minds equipped for the dynamics of the real world, minds ready to shape a complex, uncertain, and interconnected future.

Chapter 7

Conclusions

Not only in research, but also in the everyday world of politics and economics, we would all be better off if more people realized that simple nonlinear systems do not necessarily possess simple dynamical properties[201].

— Robert May

One key idea worth emphasizing is that, although the physics of complex systems currently occupies a central place in contemporary scientific research, the concept of complexity itself has a much longer intellectual history. Its origins can be traced back to the late 19th or early 20th century, and since then it has evolved through various phases, shaped by contributions from multiple disciplines, to inform the understanding we have today. Nonetheless, the development and trajectory of complexity science in the 21st century remain open-ended and full of possibilities. As new theoretical frameworks and computational tools emerge, the field continues to expand its reach, offering novel ways to approach some of the most challenging problems in science and society.

---

[201]Robert M. May. Simple Mathematical Models with very Complicated Dynamics. Nature 261, 459 (1976).

The concept of emergence, in contrast to the traditional notion of reductionism, is one of the foundational principles in the physics of complex systems. Ideas related to emergence can be traced back to the early development of thermodynamics, where collective behaviors arise that are not evident from the properties of individual components. This concept continues to appear in various phenomena studied across the physical sciences.

Notably, the emergence of order from apparent disorder is exemplified by concepts such as chaos and fractals, which have served as catalysts for the development of many ideas central to complexity science. These frameworks illustrate how simple rules can give rise to intricate and often unpredictable patterns, hallmarks of complex behavior.

In this context, interdisciplinarity becomes critically important. As previously noted, many of the basic ideas associated with complexity serve not only to deepen our understanding of individual systems but also to bridge different disciplines. Complexity science, by its very nature, encourages the integration of knowledge and the breakdown of traditional disciplinary boundaries, offering a unified perspective on the interconnected nature of the systems that shape our world.

Throughout this discussion, it is important to acknowledge that many of the ideas presented here have long been present, sometimes implicitly, in the thought and work of numerous physicists, both past and present. Many of these scientists, including several Nobel Prize laureates, have demonstrated a remarkable openness to exploring the complexities of life and nature, often venturing beyond the boundaries of traditional disciplinary frameworks.

In recent years, numerous scientists have made significant contributions to the advancement of chaos theory and the broader science of complexity. A landmark moment for the field came in 2003, when the prestigious Japan Prize, awarded annually by the Japanese government through the Japan Prize Foundation, was dedicated to the theme of Science and Technology of Complexity. That year, the prize was jointly awarded to Benoit Mandelbrot, for his pioneering work on fractals, and James A. Yorke, for his foundational contributions to chaos theory.

This recognition marked a milestone for the scientific community working in complexity-related fields. It was the first time that an award of such international stature formally honored researchers whose work lies at the heart of nonlinear dynamics, chaos, and complex systems. The award not only acknowledged the scientific merit of their contributions but also helped to elevate the visibility and legitimacy of complexity science within the broader landscape of scientific research.

Building on the efforts of numerous scientists, as previously noted, the entire field of research encompassing nonlinear dynamics, chaos theory, and complexity continues to evolve, exerting a growing influence across a wide range of disciplines. With the ongoing development of new methods and the emergence of novel theoretical frameworks, this area of study holds great promise for the future, offering fresh perspectives and tools for addressing some of the most intricate challenges in science and society.

In his influential essay *More is Different*, physicist Philip Anderson challenged the traditional hierarchical view of scientific research, which placed the highest value on discovering fundamental laws at the most elementary levels of nature. Anderson argued instead for the importance of emergence, asserting that the properties of a system at a given scale cannot be fully predicted or understood solely by analyzing the laws governing its constituent parts at lower scales. This perspective had a profound impact on the development of complexity science, emphasizing that higher-level phenomena, such as consciousness, chemical reactions, biological systems, and social behavior, are not mere extensions or applications of physics, but possess their own organizing principles and laws.

Anderson's constructivist approach marked a significant intellectual shift, encouraging scientists to appreciate complex systems in their own right and to develop new methods suited to their study. His insights laid the groundwork for a broader, interdisciplinary framework that recognizes the autonomy and richness of emergent phenomena across diverse scientific disciplines.

In his speech *Tribute to Academic Seniority 2021*,[202] Prof. Pedro García Barreno reflected on the limitations of rigid disciplinary frameworks and emphasized the importance of broader intellectual integration. He remarked:

---

[202]Pedro R. García Barreno. A personal vision. Tribute to Academic Seniority 2021. Institute of Spain, Madrid. December 14, 2021. (in Spanish)

"The cost of the disciplinary approach is that it restricts the scope of our questions and numerous extradisciplinary ideas that contribute to the progress of the cultural whole are lost. We are in a period of transscience or convergence of knowledge, an expression that recalls Goethe's Divan and that recognizes the value of approximation, of the synthesis of knowledge as an institutional priority, today through exponential technologies. Society and the Academy must wake up to the full implication of this reality. Great ideas are often characterized by considerable generality. The bigger the problems, the bigger the opportunities. The mission of the Academies is to find them."

These reflections resonate strongly with the current scientific and academic landscape, underscoring the need for intellectual openness, synthesis, and a commitment to addressing large-scale problems through the convergence of diverse forms of knowledge.

Establishing boundaries around a traditional scientific discipline is relatively straightforward; its limits, methodologies, and content are often well-defined and historically grounded. However, when dealing with a discipline that is still evolving, such as complexity science or the study of nonlinear phenomena, the situation becomes far more nuanced. Rather than viewing this as a problem, it should be seen as a unique window of opportunity. The dynamic and open-ended nature of these emerging fields invites exploration, innovation, and the possibility of redefining how we understand and integrate knowledge across disciplines. Complexity and nonlinearity, in particular, offer fertile ground for new ideas that challenge conventional boundaries and foster interdisciplinary collaboration.

Fig. 7.1   A flourishing and promising future for the science of complexity and nonlinearity.

I would like to conclude by sharing a message of optimism and hope. Despite the challenges and obstacles we have discussed, and in many ways, because of them, I firmly believe that the future of the discipline we have explored is both flourishing and full of promise. The very nature and youth of complexity science and the study of nonlinearity open up vast and exciting frontiers for discovery. Much work remains to be done. There are still many objectives to pursue, many challenges to confront, and countless opportunities to deepen our understanding of the world. It is precisely this open horizon that inspires confidence and enthusiasm for what lies ahead.

Appendix A

Response of Prof. Jesús María Sanz Serna

Excellencies Messrs. Academicians,

Ladies and Gentlemen,

It is a great satisfaction for me to be able, on behalf of the Academy, to welcome Professor Sanjuán as a new full academician. Speaking for all the members of the corporation, I hope that his scientific career continues to bring him the success that he deserves.

Born in León in 1959, Miguel Ángel Fernández Sanjuán[203] studied Physics at the University of Valladolid and received his doctorate at the National University at a Distance (UNED) under the direction of Professor Manuel García Velarde. After passing through the Polytechnic University of Madrid, he has been working at the Rey Juan Carlos University since 1997, where he has been the driving force in the creation and development of the Department of Physics, directs a strong research group and has designed an interdisciplinary degree in which, as happens in this Academy, integrates mathematics,

---

[203]NA: Full Spanish legal name of Miguel A.F. Sanjuán.

physics, chemistry, geology and biology. He has dedicated particular care and attention to the training of new researchers, as shown by the twenty theses already supervised and the very numerous post-doctoral researchers of whom he has been or is a mentor.

Among the traits of Professor Sanjuán's scientific personality I would highlight two. First of all, his almost universal curiosity. His avid interest in all kinds of issues saves him from hyperspecialization, a phenomenon that, if it enhances productivity in the short term, perhaps ends up compromising scientific progress and certainly deprives the researcher of the joy of true creation. The volume and variety of Professor Sanjuán's readings are truly exceptional, as are the number and diversity of the people with whom he has established scientific ties.

The second trait I would like to highlight is his enthusiasm for approaching, with optimism and energy, all kinds of new undertakings, tasks, and projects. Thus, our Academy has been fortunate to witness how since his election as corresponding in 2015, Professor Sanjuán has not only attended the sessions regularly, but has played an extremely active role in the life of this house, very especially in the preparation of the general conferences and in the international relations commission. Without a doubt, he has been a model corresponding academician and now as a full member he will provide the corporation with the multiple and valuable services that all its members owe to the Academy and without which it would languish no matter how great the value of the academicians. I will mention *en passant* that he is also a member of the Academia Europaea.

A tireless traveler, his work has taken Professor Sanjuán, in addition to other more predictable destinations, to Lithuania (of whose Academy of Sciences he is a foreign member), Serbia (for whose Academy of Nonlinear Sciences he has just been elected), Latvia, Poland, Russia, Ukraine, Mexico, Bolivia, Peru, Colombia, Chile, Brazil, Australia, South Africa, Dubai, Cameroon, Thailand, Turkey, Taiwan, and many others. Of particular importance are his relations with India, China and Japan. In India are the co-authors of two of his many books, one widely cited on nonlinear resonances and another recent and curious one on the physics of toys. He has spent considerable time in China, where he has received important recognitions. But, without a doubt, the place of honor among Sanjuán's scientific links belongs to Professor James Yorke of the University of Maryland, founder of chaos theory and foreign member of our Academy, with whom he maintains a very close professional and personal relationship. For this reason, I am sure that he received with great satisfaction, at the International Conference on Mathematical Analysis and Applications to Science and Engineering, the first James Yorke Award, for "breakthrough achievements in nonlinear dynamics and chaos." Two years earlier, Sanjuán had obtained the Chieh-Su Hsu Award, in nonlinear dynamics and control.

To some extent, the field of physics of the discourse we have heard, Nonlinear Dynamics, Chaos Theory and Complex Systems, could be seen as a way of transcending a venerable scientific tradition whose origin we can trace to the 1687 publication of the *Principia* of Newton. The *Principia* are without a doubt both the foundation of all physics and one of the most decisive bases of modern thought. For now, the *Principia* establishes the idea of what Newton himself called laws, that is, rules to which Nature, which classical culture had seen as capricious, inexorably subjects its operation. But, and this is what interests us now, the success of the *Principia* is also explained because Newton showed how the mathematical analysis of the physical laws of dynamics leads to being able to predict with astonishing precision the details of a plethora of phenomena. Newton, despite having devised differential and integral calculus at a very young age, always showed a preference for demonstrating its results through geometric reasoning that takes Euclid's Elements as a model. It is up to Euler, fifty years after the Englishman, to explicitly recognize that the second law of the *Principia*, force equal to mass times acceleration, is a differential equation, the resolution of which for the specific mechanical phenomenon being studied leads to its knowledge. In this way, the resolution, or as they say in mathematics, the integration, of the systems of differential equations is the key that is expected to allow us to open the mystery of the movements, perhaps complex, of bodies, in a Nature governed by laws. This idea, which first appeared in Newtonian celestial mechanics, has been extended to all types of areas, including, among many others, the mechanics of continuous media (fluids or elastic solids), electromagnetism, quantum mechanics, the kinetics of chemical reactions, the quantification of interactions between species in ecology, epidemiology (as the recent COVID-19 crisis has reminded us), or geological studies of our planet's past.

If the success of the *Principia* would not have been possible without Newton's solution of the relevant equations, many subsequent scientific milestones are also due to the completion of the integration of one or another differential equations and this explains the continued efforts in that direction. An important relatively recent case is provided by the Korteweg-de Vries and cubic Schroedinger equations, with important consequences for the study of optical fiber and several other issues. But as admirable as the successes of all the sciences have been based on that approach that dates back to the *Principia*, the ultimate truth is that most nonlinear differential equations are not integrable, they cannot be solved no matter how much daring or ingenuity we bring to the task. Unfortunately this fact has been hidden in teaching until not long ago, limiting interest to typically linear phenomena where integration is possible. Faced with this concealment, the study of nonlinear dynamics opened science to a series of fascinating and previously unknown phenomena that have been admirably glossed by Professor Sanjuán. As is usually the case in these issues, the same nonlinear phenomenon can appear in very different applications, and this fact has been combined with Sanjuán's curiosity to give his research a notable interdisciplinary dimension.

Although some of the most relevant ideas of nonlinear science were already understood, or at least anticipated, by Maxwell or Poincaré, it is fair to say that it has not been possible to make substantive advances in the nonlinear world until computers have become available that allow phenomena to be simulated when integration is not possible. After all, the first computers were designed and built for the sole purpose of solving nonlinear problems. The interaction between numerical simulation and nonlinear science is perfectly glossed in the discourse we have heard (Fermi-Pasta-Ulam-Tsingou experiment, Lorenz system, Feigenbaum cascade in the logistic equation, etc.)

After several decades of spectacular advances, nonlinear science is far from being exhausted, and I have no doubt that in this very hall, we will continue to hear interesting inaugural speeches in the decades to come, analyzing the dynamics of all kinds of nonlinear phenomena across the various sciences.

Another prominent aspect of the discourse has been the criticism of reductionism of thinking that all science is enclosed in the formulation of a few fundamental laws, understanding the term foundation in its etymological sense of foundation or basement. The speech, despite its relative brevity, amply explains the great role played by the *emergence* of new phenomena linked to the idea of *complexity*. Once again we can see in reductionism a reading, erroneous because it is exaggerated, of the physics of the *Principia*.

I will conclude with a short digression. In recent months it has been shown that, by feeding only data into a deep neural network, it is possible to make weather forecasts with the same level of accuracy and reliability as the usual ones, based on knowledge of the laws of the motion of the atmosphere and the numerical simulation of the corresponding nonlinear dynamics, in addition to statistical tools.

Is the scientific paradigm founded on laws, inaugurated with the *Principia*, now under threat? Will future physics be based on the mere handling of data by computers rather than on the idea of law? Will the paradigm that finally triumphs combine laws with data? Without a doubt we are at a point of great interest and science in the coming years or decades will make advances in these matters that are now unforeseeable. The advances in the last eighty years have also been great, which Professor Sanjuán has admirably summarized in a speech for which I thank you on behalf of all of you.

I said.

Appendix B

Interdisciplinary Research in Nonlinear Dynamics, Chaos and Complex Systems

B.1   Dynamical Systems and Fractal Structures

- J Aguirre, RL Viana, and MAF Sanjuán. Fractal structures in nonlinear dynamics. Reviews of Modern Physics 81 (1), 333 (2009).

- A Daza, A Wagemakers, B Georgeot, D Guéry-Odelin, and MAF Sanjuán. Basin entropy: a new tool to analyze uncertainty in dynamical systems. Sci. Rep. 6 (1), 31416 (2016).

- J Aguirre, JC Vallejo, and MAF Sanjuán. Wada basins and chaotic invariant sets in the Hénon-Heiles system. Phys. Rev. E 64 (6), 066208 (2001).

- J Aguirre and MAF Sanjuán. Unpredictable behavior in the Duffing oscillator: Wada basins. Physica D: Nonlinear Phenomena 17, 41-51 (2002).

- A Daza, A Wagemakers, MAF Sanjuán, and JA Yorke. Testing for basins of Wada. Sci. Rep. 5 (1), 16579 (2015).

- M Coccolo, BB Zhu, MAF Sanjuán, and JM Sanz-Serna. Bogdanov–Takens resonance in time-delayed systems. Nonlinear Dynamics 91, 1939-1947 (2018).

- A Daza, A Wagemakers, and MAF Sanjuán. Multistability and unpredictability. Physics Today 77 (11), 44–50 (2024).

B.2 Celestial Mechanics and Astrophysics

- A Daza, JO Shipley, SR Dolan, and MAF Sanjuán. Wada structures in a binary black hole system. Phys. Rev. D 98, 084050 (2018).
- JC Vallejo and MAF Sanjuán. Predictability of Chaotic Dynamics. A finite-time Lyapunov exponent approach. (Springer Series in Synergetics), Springer, Cham, 2019. Second Edition.
- JC Vallejo and MAF Sanjuán. The Forecast of Predictability in Galactic Potentials. Mon. Notices Royal Astron. Soc. 447, 3797-3811 (2015).
- JM Muñoz, A Wagemakers, and MAF Sanjuán. Planetary influences on the solar cycle: A nonlinear dynamics approach. Chaos 33(12), 123102 (2023).
- A. Meseguer, JC Vallejo, JM Seoane, F Marqués and MAF Sanjuán. Escaping dynamics of relativistic protons in the Earth's magnetosphere. Phys. Rev. E 111, 044216 (2025).

B.3 Applications in Physics

- SK Joseph, LY Chew, and MAF Sanjuán. Impact of quantum–classical correspondence on entanglement enhancement by single-mode squeezing. Phys. Lett. A 378(35), 2603-2610 (2014).

- A Daza, B Georgeot, D Guéry-Odelin, A Wagemakers, and MAF Sanjuán. Chaotic dynamics and fractal structures in experiments with cold atoms. Phys. Rev. A 95 (1), 013629 (2017).

- JM Seoane and MAF Sanjuán. New developments in classical chaotic scattering. Rep. Prog. Phys. 76(1), 016001 (2012).

- EC da Silva, IL Caldas, RL Viana, and MAF Sanjuán. Escape patterns, magnetic footprints, and homoclinic tangles due to ergodic magnetic limiters. Physics of Plasmas 9, 4917-4928 (2002).

- JSE Portela, IEL Caldas, RL Viana, and MAF Sanjuan. Fractal and Wada exit basin boundaries in tokamaks. Int J Bifurcat Chaos 17, 4067-4079 (2007).

- RL Viana, EC Da Silva, T Kroetz, IL Caldas, M Roberto, and MAF Sanjuán. Fractal structures in nonlinear plasma physics. Phil. Trans. R. Soc. A. 369, 371–395 (2011).

- A Wagemakers, A Hartikainen, A Daza, E Räsänen, and Miguel A. F. Sanjuán. Chaotic dynamics creates and destroys branched flow. Phys. Rev. E 111, 014214 (2025).

B.4   Control of Chaotic Systems

- MAF Sanjuán and C Grebogi, Recent Progress in Controlling Chaos (World Scientific, Singapore, 2010).

- S Zambrano, MAF Sanjuán, and JA Yorke. Partial control of chaotic systems. Phys. Rev. E 77, 055201(R) (2008).

- J Sabuco, MAF Sanjuán, and JA Yorke. Dynamics of partial control. Chaos 22, 047507 (2012).

- R Capeáns, J Sabuco, MAF Sanjuán, and JA Yorke. Partially controlling transient chaos in the Lorenz equations. Phil. Trans. R. Soc. A 375, 20160211 (2017).

- SK Joseph, IP Mariño, and MAF Sanjuán. Effect of the phase on the dynamics of a disturbed bouncing ball system. Commun Nonlinear Sci Numer Simul 17, 3279-3286 (2012).

- S Roy, M Coccolo, and MAF Sanjuán. Parametric autoresonance with time-delayed control. Phys. Rev. E 111(1), 014225 (2025).

B.5   Noise and Stochastic Phenomena

- AR Nieto, JM Seoane, and MAF Sanjuán. Final state sensitivity in noisy chaotic scattering. Chaos Solitons Fractals 150, 111181 (2021).

- J Cantisán, JM Seoane, and MAF Sanjuán. Stochastic resetting in the Kramers problem: A Monte Carlo approach. Chaos Solitons Fractals 152, 111342 (2021).

B.6   Machine Learning, AI and Evolutionary Game Theory

- MAF Sanjuán. Artificial intelligence, chaos, prediction and understanding in science. Int J Bifurcat Chaos 31 (11), 2150173 (2021).

- D Valle, A Wagemakers, A Daza, and MAF Sanjuán. Characterization of fractal basins using deep convolutional neural networks. Int J Bifurcat Chaos 32, 2250200 (2022).

- G Alfaro, and MAF Sanjuán. Hamming distance as a measure of spatial chaos in evolutionary games. Phys. Rev. E 109, 014203 (2024).

- D Valle, A Wagemakers, and MAF Sanjuán. Deep Learning-based Analysis of Basins of Attraction. Chaos 34, 033105 (2024).

- D Valle, R Capeáns, A Wagemakers, and MAF Sanjuán. AI-driven control of chaos: A transformer-based approach for dynamical systems. Commun Nonlinear Sci Numer Simulat 151, 109085 (2025).

B.7   Nonlinear Mechanics

- H Cao, JM Seoane, and MAF Sanjuán. Symmetry-breaking analysis for the general Helmholtz–Duffing oscillator. Chaos Solitons Fractals 34, 197-212 (2007).

- JH Yang, MAF Sanjuán, and HG Liu. Vibrational subharmonic and superharmonic resonances. Commun Nonlinear Sci Numer Simulat 30 (1-3), 362-372 (2016).

- AA Zaikin, L López, JP Baltanás, J Kurths, and MAF Sanjuán. Vibrational resonance in a noise-induced structure. Phys. Rev. E 66 (1), 011106 (2002).

- C Jeevarathinam, S Rajasekar, and MAF Sanjuán. Theory and numericals of vibrational resonance in Duffing oscillators with time-delayed feedback. Phys. Rev. E 83 (6), 066205 (2011).

- S Rajasekar, K Abirami, and MAF Sanjuán. Novel vibrational resonance in multistable systems. Chaos 21, 033106 (2011).

- S Rajasekar and MAF Sanjuan. Nonlinear resonances (Springer Series in Synergetics), (Springer, Cham, 2016).

- JH Yang, S Rajasekar, and MAF Sanjuan. Vibrational Resonance: A Review. Physics Reports 1067, 1-62 (2024).

B.8   Biological Phenomena

- B Ibarz, JM Casado, and MAF Sanjuán. Map-based models in neural dynamics. Physics Reports 501 (1-2), 1-74 (2011).

- AG López, J Sabuco, JM Seoane, J Duarte, C Januário, and MAF Sanjuán. Avoiding healthy cells extinction in a cancer model. J. Theor. Biol. 349, 74-81 (2014).

- AG López, JM Seoane, and MAF Sanjuán. A validated mathematical model of tumor growth including tumor–host interaction, cell-mediated immune response and chemotherapy. Bull. Math. Biol. 76, 2884-2906 (2014).

- C Jeevarathinam, S Rajasekar, and MAF Sanjuán. Vibrational resonance in groundwater-dependent plant ecosystems. Ecological Complexity 15, 33-42 (2013).

- R Capeáns, J Sabuco, and MAF Sanjuán. When less is more: Partial control to avoid extinction of predators in an ecological model. Ecological Complexity 19, 1-8 (2014).

- A Wagemakers, JM Buldú, J García-Ojalvo, and MAF Sanjuán. Synchronization of electronic genetic networks. Chaos 16, 013127 (2006).

- Irina Bashkirtseva, Lev Ryashko, Javier Used, Jesús M. Seoane, and Miguel AF Sanjuán. Noise-induced complex dynamics and synchronization in the map-based Chialvo neuron model. Commun Nonlinear Sci Numer Simulat 116, 106867 (2023).

- J Duarte, C Januario, N Martins, J Seoane, and MAF Sanjuán. Controlling infectious diseases: the decisive phase effect on a seasonal vaccination strategy. Int J Bifurcat Chaos 31, 2130044 (2021).

B.9   Fundamentals of Nonlinear Dynamics and Chaos Theory

- MAF Sanjuán, J Kennedy, E Ott, and JA Yorke. Indecomposable continua and the characterization of strange sets in nonlinear dynamics. Phys. Rev. Lett. 78, 1892 (1997).

- MAF Sanjuán, J Kennedy, C Grebogi, and JA Yorke. Indecomposable continuous in dynamical systems with noise: Fluid flow past an array of cylinders. Chaos 7, 125-138 (1997).

- Y Saiki, MAF Sanjuán, and JA Yorke. Low-dimensional paradigms for high-dimensional hetero-chaos. Chaos 28, 103110 (2018).

About the Author

 **MIGUEL A. F. SANJUÁN** received his bachelor's degree in physics from the University of Valladolid, Spain, in 1981, where he was awarded the Outstanding Graduation Honor for Undergraduate Studies. He earned his Ph.D. in Physics from the National University of Distance Education (UNED), Madrid, Spain, in 1990, specializing in Nonlinear Dynamics and Chaos.

He is currently a Professor of Physics at Universidad Rey Juan Carlos, Madrid, Spain, and Director of the Research Group in Nonlinear Dynamics, Chaos, and Complex Systems. He is the author of numerous research articles and several books in the field. He serves on the editorial boards of several international journals in Nonlinear Dynamics and Chaos and is the Editor-in-Chief of the Journal of Applied Nonlinear Dynamics.

He has delivered invited lectures at universities across Spain, Europe, the USA, Canada, China, Japan, India, Australia, South America, and Africa. His international academic collaborations include positions as Visiting Research Associate at the Institute for Physical Sciences and Technology, University of Maryland; Visiting Research Professor at Beijing Jiaotong University; Visiting Professor at Kaunas University of Technology, Lithuania; and Guest Professor at Lanzhou University and Zigong University, China. He was also a JSPS Fellow of the Japan Society for the Promotion of Science at the University of Tokyo.

In 2017, he was a Fulbright Visiting Research Scholar at the University of Maryland. He was the first recipient of the Chieh-Su Hsu Award (2020) for distinguished contributions to Nonlinear Dynamics and Control, and in 2022, he received the James Yorke Award for breakthrough achievements in Nonlinear Dynamics and Chaos.

He holds the title of Honorary Professor at both Sichuan University of Science and Technology (Zigong, China) and Huaqiao University (Xiamen, China). He currently serves as Editor General of the Spanish Royal Physics Society (RSEF) and is a member of the EPJ Scientific Advisory Committee of the European Physical Society (EPS).

He has taught courses on Nonlinear Dynamics, Chaos, and Complex Systems at Universidad Rey Juan Carlos (Madrid, Spain), Kaunas University of Technology (Kaunas, Lithuania), Beijing Jiaotong University, and Peking University (Beijing, China).

He is a Full Member of the Royal Academy of Sciences of Spain, a Foreign Member of the Lithuanian Academy of Sciences, a Foreign Member of the Serbian Academy of Nonlinear Sciences, and a Regular Member of the Academia Europaea — The Academy of Europe.

Made in the USA
Monee, IL
07 July 2026

b274963d-bc35-407c-a2cf-3ee4b68912e4R01